Introduction to AutoC

Introduction to AutoCAD 2005
2D and 3D Design

Alf Yarwood

AMSTERDAM • BOSTON • HEIDELBERG • LONDON
NEW YORK • OXFORD • PARIS • SAN DIEGO
SAN FRANCISCO • SINGAPORE • SYDNEY • TOKYO

Newnes is an imprint of Elsevier

ELSEVIER

Newnes

Newnes
An imprint of Elsevier
Linacre House, Jordan Hill, Oxford OX2 8DP
30 Corporate Drive, Burlington, MA 01803

First published 2005

British Library Cataloguing in Publication Data
A catalogue record for this book is available from the British Library

Library of Congress Cataloguing in Publication Data
A catalogue record for this book is available from the Library of Congress

ISBN   0 7506 67214

For information on all Newnes publications
visit our website at http://books.elsevier.com

Typeset by Integra Software Services Pvt. Ltd, Pondicherry, India
www.integra-india.com
Printed and bound in Great Britain

Working together to grow
libraries in developing countries

www.elsevier.com | www.bookaid.org | www.sabre.org

ELSEVIER   BOOK AID International   Sabre Foundation

Contents

Preface

The purpose of writing this book is to produce a text suitable for those in Further and/or Higher Education who are required to learn how to use the computer-aided design (CAD) software package AutoCAD® 2005. The book is also suitable for those in industry wishing to learn how to construct technical drawings with the aid of AutoCAD 2005 and those who, having used previous releases of AutoCAD, wish to update their skills in the use of AutoCAD.

The chapters dealing with two-dimensional (2D) drawing will also be suitable for those wishing to learn how to use AutoCAD LT 2005 – the 2D version of this latest release of AutoCAD.

Many readers using AutoCAD 2002 or 2004 will find the book's contents largely suitable for use with those versions of AutoCAD, although AutoCAD 2005 has enhancements over both AutoCAD 2002 and 2004 (see Chapter 22).

The contents of the book are basically a graded course of work, consisting of chapters giving explanations and examples of methods of constructions, followed by exercises which allow the reader to practise what has been learned in each chapter. The first twelve chapters are concerned with constructing technical drawing in two dimensions (2D). These are followed by chapters detailing the construction of three-dimensional (3D) solid model drawing and rendering. The two final chapters describe the Internet tools of AutoCAD 2005 and the place of AutoCAD in the design process. The book finishes with four appendices: printing and plotting; a list of tools with their abbreviations; a list of some of the set variables upon which AutoCAD 2005 is based; and a final appendix describing common computing terms.

AutoCAD 2005 is a very complex CAD software package. A book of this size cannot possibly cover the complexities of all the methods for constructing 2D and 3D drawings available when working with AutoCAD 2005. However, it is hoped that by the time the reader has worked through the contents of the book, they will be sufficiently skilled with methods of producing drawing with the software, will be able to go on to more advanced constructions with its use and will have gained an interest in the more advanced possibilities available when using AutoCAD.

Alf Yarwood
Salisbury 2005

Registered Trademarks

Autodesk® and AutoCAD® are registered trademarks of Autodesk, Inc., in the USA and/or other countries. All other brand names, product names, or trademarks belong to their respective holders.

Windows® is a registered trademark of the Microsoft Corporation.

Alf Yarwood is an Autodesk authorised publisher and a member of the Autodesk Developer Network.

CHAPTER 1

Introducing AutoCAD 2005

Aim of this chapter

The contents of this chapter are designed to introduce features of the AutoCAD 2005 window and methods of operating AutoCAD 2005.

Opening AutoCAD 2005

Fig. 1.1 The **AutoCAD 2005** shortcut icon on the Windows desktop

AutoCAD 2005 is designed to work in a Windows operating system. In general to open AutoCAD 2005 either *double-click* on the **AutoCAD 2005** shortcut on the Windows desktop (Fig. 1.1), or *right-click* on the icon, followed by a *left-click* on **Open** in the menu which then appears (Fig. 1.2).

Fig. 1.2 The *right-click* menu which appears from the shortcut icon

When working in education or in industry computers which may be configured to allow other methods of opening AutoCAD, such as a list appearing on the computer in use when the computer is switched on, from which the operator can select the program they wish to use.

When AutoCAD 2005 in opened a window appears (Fig. 1.3). Usually the toolbars are in the positions as indicated in the Fig. 1.3. In particular the toolbars in most common use are:

Standard toolbar (Fig. 1.4): *Docked* at the top of the AutoCAD window under the **Menu bar**.

Draw toolbar: *Docked* against the left-hand side of the AutoCAD window.

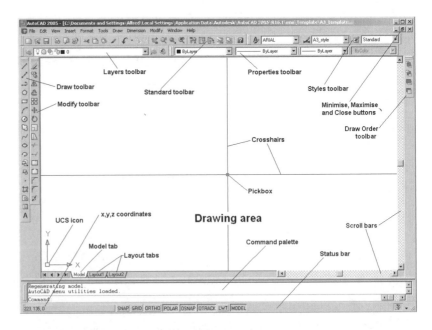

Fig. 1.3 The AutoCAD 2005
window shown with its various
parts

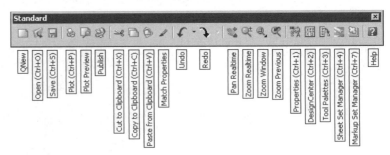

Fig. 1.4 The tool icons in the
Standard toolbar

Modify toolbar: *Docked* against the **Draw** toolbar.
Layers toolbar: *Docked* under the **Standard** toolbar.
Properties toolbar: *Docked* to the right of the **Layers** toolbar.
Styles toolbar: *Docked* to the right of the **Properties** toolbar.
Draw Order toolbar: May or may not be as shown.
Command palette: Can be *dragged* from its position as shown into the
    AutoCAD drawing area, when it can be seen as a palette (Fig. 1.5).

Fig. 1.5 The command palette
when *dragged* from its position
at the bottom of the AutoCAD
window

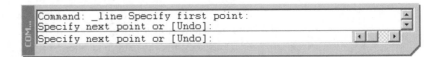

Menu bar and **menus**: The **menu bar** is situated just under the **title
bar** and contains the names of menus from which tools and
commands can be selected. Fig. 1.6 shows the drop-down menu
which appears with a *left-click* on the name **View** in the **menu bar**.

A *left-click* on the name **3D Views** in the drop-down menu brings a sub-menu on screen, from which other sub-menus can be selected if required.

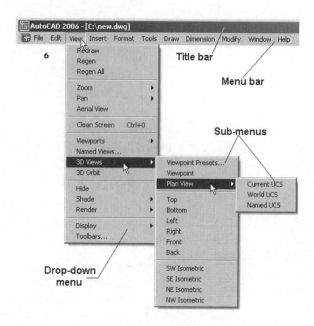

Fig. 1.6 Menus and sub-menus

The mouse as a digitiser

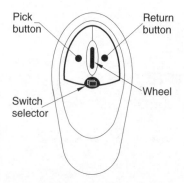

Fig. 1.7 A two-button cordless mouse

Most operators using AutoCAD use a two-button mouse as the digitiser. There are other forms of digitiser which may be used such as pucks with tablets, a three-button mouse, etc. Fig. 1.7 shows a cordless two-button mouse which, in addition to its two buttons, has a wheel and a switch selection button.

To operate this mouse pressing the **Pick button** is a *left-click*. Pressing the **Return button** is a *right-click*. Pressing the **Return button** usually, but not always, has the same result as pressing the **Enter** key of the computer's keyboard.

When the **Wheel** is pressed down an icon appears in the AutoCAD window (Fig. 1.8). When the icon is on screen, moving the mouse pans a drawing on screen. Moving the wheel forwards enlarges (zooms in) the drawing on screen. Move the wheel backwards and a drawing reduces in size on the screen (zooms out).

Press the **Switch selector** and a window appears (Fig. 1.9) showing which applications are loaded on the computer in the Windows multi-tasking system. *Left-click* on any one of the applications named in the window and that application becomes current.

The pick box at the intersection of the cursor hairs moves with the cursor hairs in response to movements of the mouse. The AutoCAD window as shown in Fig. 1.3 includes cursor hairs which stretch across the drawing in both horizontal and vertical directions. Some operators prefer cursors

Fig. 1.8 The icon which appears when the **Wheel** is pressed

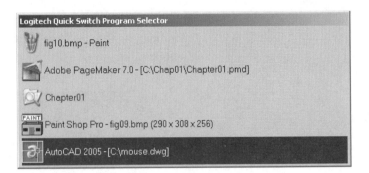

Fig. 1.9 The **Switch Program Selector** window

hairs to be shorter. The length of the cursor hairs can be adjusted in the **Options** dialog (page 7).

Palettes

Two palettes which may be frequently used are the **DesignCenter** palette and the **Properties** palette. These can be called to screen from icons in the **Standard** toolbar as shown in Figs 1.10 and 1.11.

DesignCenter palette: Fig. 1.12 shows the palette showing the **Block** drawings of metric fasteners from an AutoCAD directory **DesignCenter** from which the drawing file **Fasteners – Metric.dwg** has been selected. A fastener block drawing can be *dragged* from the **DesignCenter** for inclusion in a drawing under construction.

Fig. 1.10 The **DesignCenter** icon in the **Standard** toolbar

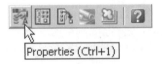

Fig. 1.11 The **Properties** icon in the **Standard** toolbar

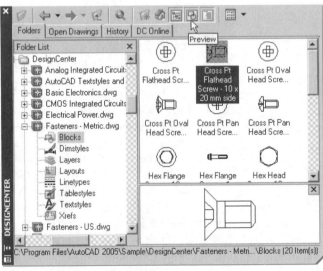

Fig. 1.12 The **DesignCenter** palette

Properties palette: Fig. 1.13 shows the **Properties** palette in which the general and geometrical features of a selected polyline are shown. The polyline can be changed by the *entering* of new figures in the appropriate parts of the palette.

Fig. 1.13 The **Properties** palette

Toolbars

Tools used in the construction of drawings in AutoCAD 2005 are held in toolbars. The list of available toolbars is shown in the menu in Fig. 1.14. This menu is called to screen with a *right-click* on any toolbar already on screen. Toolbars already on screen are shown by ticks against their names in the menu. To call a new toolbar to screen *left-click* on its name in the menu.

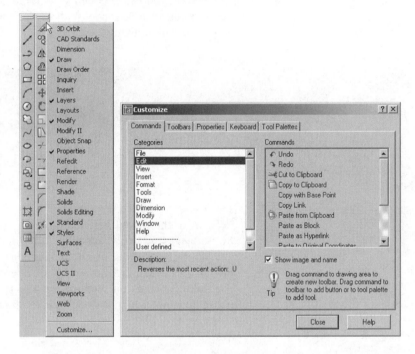

Fig. 1.14 The toolbars menu and the **Customize** dialog

A *left-click* on **Customize** . . . in the menu brings a dialog on screen by which one can construct a toolbar to suit one's own drawing methods.

When a toolbar is selected from the menu it appears on screen as shown in the top of Fig. 1.15. By *dragging* on cursors which appear at the edges of the toolbar, when the cursor hairs under mouse movement are placed in position, the shape of the toolbar can be changed. Toolbars in the drawing area of the AutoCAD window are said to be *floating*.

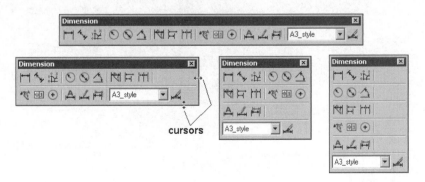

Fig. 1.15 Toolbars in the drawing area

Dialogs

Dialogs are an important feature of AutoCAD 2005. Settings can be made in many of the dialogs, files can be saved and opened, and changes can be made to variables.

Examples of the parts of dialogs are shown in Figs 1.17 to 1.19. These examples are taken from the **Select File** dialog, by which drawings which have been saved can be opened and part of the **Options** dialog in which many settings can be made to allow operators the choice of their methods of constructing drawings.

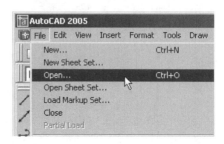

Fig. 1.16 Opening the **Select File** dialog

To open the **Select File** dialog, *click* **File** in the menu bar and in the drop-down menu which appears *click* **Open** . . . (Fig. 1.16). Note the three dots after **Open**. This means that a *click* on any such name in a drop-down menu which is followed by . . . opens a dialog to screen. The **Select File** dialog appears on screen (Fig. 1.17). Note the following parts in the dialog, many of which are common to other AutoCAD dialogs:

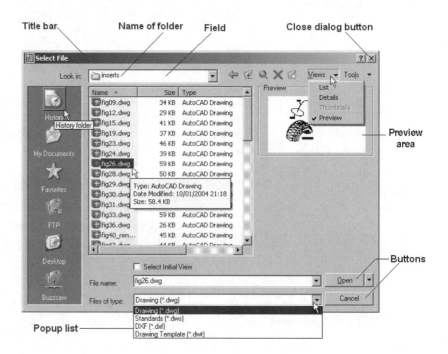

Fig. 1.17 The **Select File** dialog and its parts

Title bar: Showing the name of the dialog.

Close dialog button: Common to other dialogs.

Popup list: A *left-click* on the arrow to the right of the field brings down a popup list listing selections available in the dialog.

Buttons: A *click* on the **Open** button brings the selected drawing on screen. A *click* on the **Cancel** button closes the dialog.

Preview area: Available in some dialogs – shows a miniature of the selected drawing or other features.

To open the **Options** dialog, *right-click* in the command palette and a *right-click* menu appears (Fig. 1.18). *Click* on the name **Options** . . . and the **Options** dialog appears on screen. This is a complex dialog, only part of which is shown in Fig. 1.19.

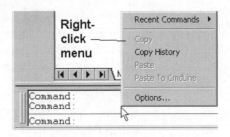

Fig. 1.18 The *right-click* menu in the command palette

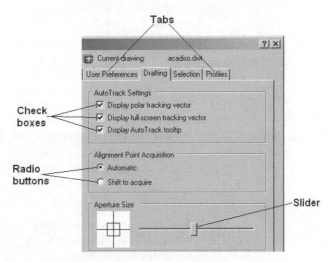

Fig. 1.19 Part of the **Options** dialog

Note the following in this dialog:

Tabs: A *click* on any of the tabs in the dialog brings a sub-dialog on screen.

Check boxes: A tick appearing in a check box indicates the function described against the box is on. No tick and the function is off. A *click* in a check box toggles between the feature being off or on.

Radio buttons: A black dot in a radio button indicates the feature described is on. No dot and the feature is off.

Slider: A slider pointer can be *dragged* to change sizes of the feature controlled by the slider.

Buttons in the status bar

A number of buttons in the status bar can be used for toggling (turning on/off) various functions when operating within AutoCAD 2005 (Fig. 1.20). A *click* on a button turns that function on, if it is off; a *click* on a button when it is off turns the function back on. Similar results can be obtained by using function keys of the computer keyboard (keys **F1** to **F10**).

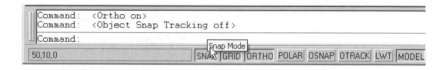

Fig. 1.20 The buttons in the status bar

SNAP: Also toggled using the **F9** key. When set on, the cursor under mouse control can only be moved in jumps from one snap point to another. See also page 34.

GRID: Also toggled using the **F7** key. When set on, a series of grid points appears in the drawing area. See also page 9.

ORTHO: Also toggled using the **F8** key. When set on, lines etc. can be drawn only vertically or horizontally.

POLAR: Also toggled using the **F10** key. When set on, a small tip appears showing the direction and length of lines etc. in degrees and units.

OSNAP: Also toggled using the **F3** key. When set on, an osnap icon appears at the cursor pick box. See also page 34.

OTRACK: When set on, lines etc. can be drawn at exact coordinate points and precise angles.

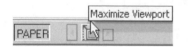

Fig. 1.21 The **Maximize Viewport** icon in the status bar

LWT: When set on, lineweights show on screen. When set off, lineweights show in only plotted/printed drawings.

MODEL: Toggles between Model Space and Paper Space. When in Paper Space an additional icon appears in the status bar (Fig. 1.21). A *click* on the icon maximises the drawing area. See also page 234.

Note

When constructing drawings in AutoCAD 2005 it is advisable to toggle between **Snap**, **Ortho**, **Osnap** and the other functions in order to make constructing easier.

The AutoCAD coordinate system

In the AutoCAD 2D coordinate system, units are measured horizontally in terms of X and vertically in terms of Y. A 2D point can be determined in terms of X,Y (in this book referred to as *x,y*). *x,y* = 0,0 is the **origin** of the system. The coordinate point *x,y* = 100,50 is 100 units to the right of the origin and 50 units above the origin. The point *x,y* = $-100,-50$ is 100 units to the left of the origin and 50 points below the origin. Fig. 1.22 shows some 2D coordinate points in the AutoCAD window.

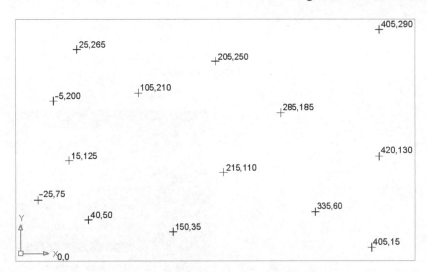

Fig. 1.22 2D coordinate points in the AutoCAD coordinate system

3D coordinates include a third coordinate (Z), in which positive Z units are towards the operator as if coming out of the monitor screen and negative Z units going away from the operator as if towards the interior of the screen. 3D coordinates are stated in terms of x,y,z. $x,y,z = 100,50,50$ is 100 units to the right of the origin, 50 units above the origin and 50 units towards the operator. A 3D model drawing as if resting on the surface of a monitor is shown in Fig. 1.23.

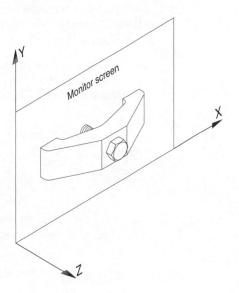

Fig. 1.23 A 3D model drawing showing the X, Y and Z coordinate directions

Drawing templates

Drawing templates are files with an extension **.dwt**. Templates are files which have been saved with predetermined settings, such as **Grid** spacing, **Snap** spacing, etc. Templates can be opened from the **Select template** dialog (see Fig. 1.17, page 6) called by *clicking* **New** ... in the **File**. An example of a template file being opened is shown in Fig. 1.24. In this example the template is opened in Paper Space and is complete with a title block and borders.

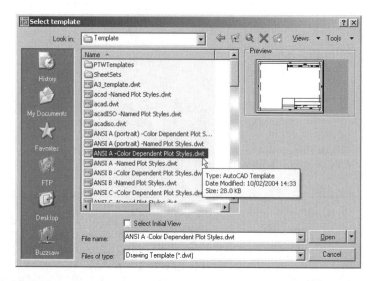

Fig. 1.24 A template selected for opening in the **Select template** dialog

When AutoCAD 2005 is used in European countries and opened, the **acadiso.dwt** template automatically appears on screen. Throughout this book drawings will usually be constructed in an adaptation of the **acadiso.dwt** template. To adapt this template:

1. In the command palette *enter* (type) **grid** followed by a *right-click* (or by pressing the **Enter** key). Then *enter* **10** in response to the prompt which appears, followed by a *right-click*. (Fig. 1.25).

```
Command: grid
Specify grid spacing(X) or [ON/OFF/Snap/Aspect] <0.5>: 10
Command:
```

Fig. 1.25 Setting **Grid** to **10**

2. In the command palette *enter* **snap** followed by *right-click*. Then *enter* **5** followed by a *right-click* (Fig. 1.26).

```
Command: snap
Specify snap spacing or [ON/OFF/Aspect/Rotate/Style/Type] <0.5>: 5
Command:
```

Fig. 1.26 Setting **Snap** to **5**

3. In the command palette *enter* **limits**, followed by a *right-click. Right-click* again. Then *enter* **420,297** and *right-click* (Fig. 1.27).

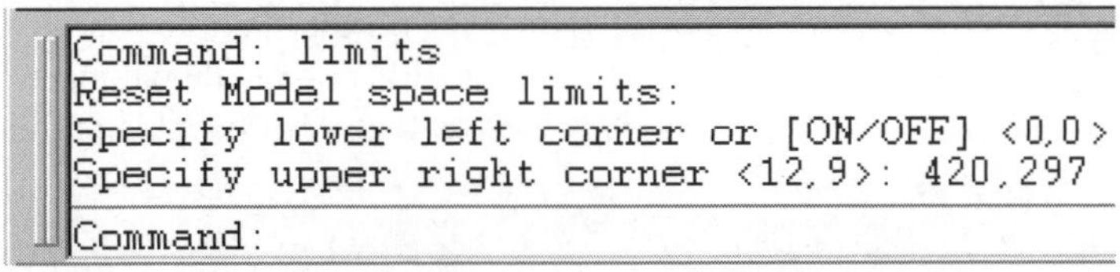

```
Command: limits
Reset Model space limits:
Specify lower left corner or [ON/OFF] <0,0>
Specify upper right corner <12,9>: 420,297
Command:
```

Fig. 1.27 Setting **Limits** to
420,297

4. In the command window *enter* **zoom** and *right-click*. Then in response to the line of prompts which appears *enter* **a** (for All) and *right-click* (Fig. 1.28).

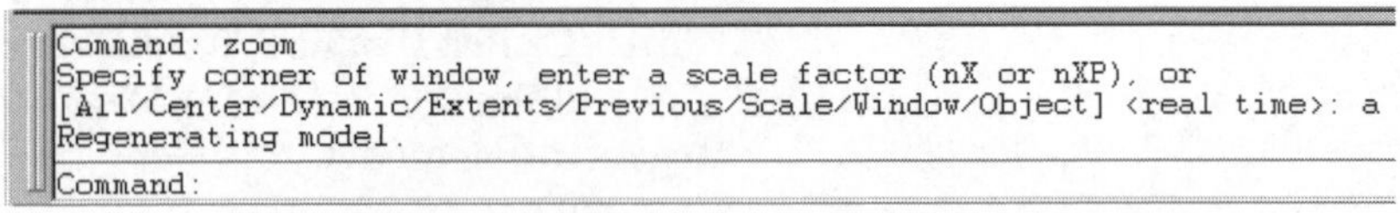

```
Command: zoom
Specify corner of window, enter a scale factor (nX or nXP), or
[All/Center/Dynamic/Extents/Previous/Scale/Window/Object] <real time>: a
Regenerating model.
Command:
```

Fig. 1.28 **Zooming** to **All**

5. In the command palette *enter* **units** and *right-click*. The **Units** dialog appears (Fig. 1.29). In the **Precision** popup list of the **Length** area of the dialog, *click* on **0** and then *click* the **OK** button. Note the change in the coordinate units showing in the status bar.

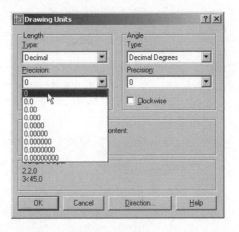

Fig. 1.29 Setting **Units** to **0**

6. *Click* **File** in the menu bar and *click* **Save As** . . . in the drop-down menu which appears. The **Save Drawing As** dialog appears. In the **Files of type** popup list select **AutoCAD Drawing Template (*.dwt)**. The templates already in AutoCAD are displayed in the dialog. *Click* on **acadiso.dwt**, followed by another *click* on the **Save** button.

Notes

1. Now when AutoCAD is opened the template saved as **acadiso.dwt** automatically loads with **Grid** set to **10**, **Snap** set to **5**, **Limits** set to **420,297** (size of an A3 sheet in millimetres), with the drawing area zoomed to these limits and with **Units** set to **0**.

2. However if other people use the computer, it is advisable to save your template to another file name, such as **my_template.dwt**.

3. Other features will be added to the template in future chapters.

Method of showing entries in the command palette

Throughout the book, where necessary, details *entered* in the command palette will be shown as follows:
At the command line:

> **Command:** *enter* **zoom** *right-click*
> **Specify corner of window, enter a scale factor (nX or nXP), or**
> **[All/Center/Dynamic/Extents/Previous/Scale/Window/Object]**
> **<real time>:** *enter* **a** (All) *right-click*
> **Regenerating model.**
> **Command:**

Note

In later examples this will be shortened to:

> **Command:** *enter* **z** *right-click*
> **[prompts]:** *enter* **a** *right-click*
> **Command:**

Notes

1. In the above *enter* means type the given letter, word or words at the **Command:** prompt.

2. *Right-click* means press the **Return** (right) button of the mouse.

Tools and tool icons

An important feature of Windows applications are icons and tips. In AutoCAD 2005, tools are shown as icons in toolbars. When the cursor is placed over a tool icon a tool tip shows with the name of the tool.

If a small arrow is included at the bottom right-hand corner of a tool icon, when the cursor is placed over the icon and the *pick* button of the mouse depressed and held, a flyout appears which includes other tool icons (Fig. 1.30). Flyouts contain icons for tools related to the tool showing the arrow in its icon.

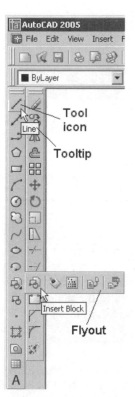

Fig. 1.30 Tool icons in the **Draw** and **Modify** toolbars and the **Line** tool tip

Revision notes

1. A *double-click* on the **AutoCAD 2005** shortcut on the Windows desktop opens the AutoCAD window.

2. Or *right-click* on the shortcut, followed by a *left-click* on **Open** in the menu which then appears.

3. The **Standard**, **Layers**, **Properties**, **Styles**, **Draw** and **Modify** toolbars usually appear *docked* against the top or sides of the AutoCAD window when it is opened.

4. A *left-click* on a menu name in the menu bar brings a drop-down menu on screen. In drop-down menus:
   (a) A small outward pointing arrow against a name means that a sub-menu will appear with a *click* on the name.
   (b) Three dots (. . .) following a name mean that a *click* on the name will bring a dialog on screen.
5. All constructions in this book involve the use of a mouse as the digitiser. When a mouse is the digitiser:
   (a) A *left-click* means pressing the left-hand button (the **Pick**) button.
   (b) A *right-click* means pressing the right-hand button (the **Return**) button.
   (c) A *double-click* means pressing the left-hand button twice is quick succession.
   (d) *Dragging* means moving the mouse until the cursor is over an item on screen, holding the left-hand button down and moving the mouse. The item moves in sympathy with the mouse movement.
   (e) To *pick* has a similar meaning to a *left-click*.
6. Palettes are a feature of AutoCAD 2005. In particular the **Command** palette, the **DesignCenter** palette and the **Properties** palette may be in frequent use.
7. Tools are shown as icons in toolbars.
8. When a tool is *picked* a tool tip describing the tool appears.
9. A toolbar menu appears with a *right-click* in any toolbar on screen.
10. Dialogs allow opening and saving of files and the setting of parameters.
11. A number of *right-click* menus are used in AutoCAD 2005.
12. A number of buttons in the status bar can be used to toggle features such as snap and grid. Function keys of the keyboard can also be used for toggling some of these functions.
13. The AutoCAD coordinate system determines the position in units of any point in the drawing area and any point in 3D space.
14. Drawings are usually constructed in templates with predetermined settings. Some templates include borders and title blocks.

CHAPTER 2

Introducing drawing

Aims of this chapter

The contents of this chapter are designed to introduce:

1. The drawing of simple outlines using the **Line**, **Circle** and **Polyline** tools from the **Draw** toolbar.
2. Drawing to snap points.
3. Drawing to absolute coordinate points.
4. Drawing to relative coordinate points.
5. Drawing using the 'tracking' method.
6. The use of the **Erase**, **Undo** and **Redo** tools.

Drawing with the Line tool

First example – Line tool (Fig. 2.3)

1. Open AutoCAD. The drawing area will show the settings of the **acadiso.dwt** template – **Limits** set to **420,297**, **Grid** set to **10**, **Snap** set to **5** and **Units** set to **0**.
2. *Left-click* on the **Line** tool at the top of the **Draw** toolbar (Fig. 2.1).

Fig. 2.1 The **Line** tool icon in the **Draw** toolbar with its tooltip

Notes

(a) The tooltip which appears when the tool icon is *clicked*.
(b) The prompt **Command:_line Specify first point** which appears in the command window at the command line (Fig. 2.2).
(c) The description of the action of the **Line** tool which appears at the left-hand end of the status bar (Fig. 2.2).

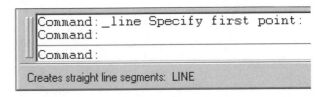

Fig. 2.2 The prompts appearing at the command line in the Command palette when **Line** is 'called'

3. Press the **F6** key. The coordinate numbers at the left-hand end of the status bar either show clearly or grey out. Make sure the coordinate numbers show clearly (black) when **F6** is pressed.

4. Make sure **Snap** is on by pressing either the **F9** key or the **SNAP** button in the status bar. **<Snap on>** will show in the Command palette.
5. Move the mouse around the drawing area. The cursors' pick box will jump from point to point at 5 unit intervals. The position of the pick box will show as coordinate numbers in the status bar.
6. Move the mouse until the coordinate numbers show **60,240,0** and press the **Pick** button of the mouse (*left-click*).
7. Move the mouse until the coordinate numbers show **260,240,0** and *left-click*.
8. Move the mouse until the coordinate numbers show **260,110,0** and *left-click*.
9. Move the mouse until the coordinate numbers show **60,110,0** and *left-click*.
10. Move the mouse until the coordinate numbers show **60,240,0** and *left-click*. Then press the **Return** button of the mouse (*right-click*).

Fig. 2.3 appears in the drawing area.

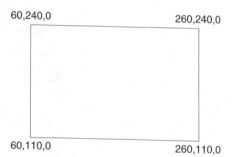

Fig. 2.3 First example – **Line** tool

Second example – Line tool (Fig. 2.6)

1. Clear the drawing from the screen with a *click* on the drawing **Close** button (Fig. 2.4). Make sure it is not the AutoCAD 2005 window button.

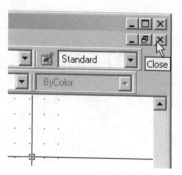

Fig. 2.4 The drawing **Close** button

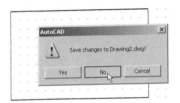

Fig. 2.5 The **AutoCAD** warning window

2. The warning window (Fig. 2.5) appears in the centre of the screen. *Click* its **No** button.
3. *Left-click* on **New** . . . in the **File** drop-down menu and from the **Select template** dialog which appears *double-click* on **acadiso.dwt**.
4. *Left-click* on the **Line** tool icon and *enter* figures as follows at each prompt of the command line sequence:

Command:_line Specify first point: *enter* **80,235** *right-click*
Specify next point or [Undo]: *enter* **275,235** *right-click*
Specify next point or [Undo]: *enter* **295,210** *right-click*
Specify next point or [Close/Undo]: *enter* **295,100** *right-click*
Specify next point or [Close/Undo]: *enter* **230,100** *right-click*
Specify next point or [Close/Undo]: *enter* **230,70** *right-click*
Specify next point or [Close/Undo]: *enter* **120,70** *right-click*
Specify next point or [Close/Undo]: *enter* **120,100** *right-click*
Specify next point or [Close/Undo]: *enter* **55,100** *right-click*
Specify next point or [Close/Undo]: *enter* **55,210** *right-click*
Specify next point or [Close/Undo]: *enter* **c** (Close) *right-click*
Command:

The result is as shown in Fig. 2.6.

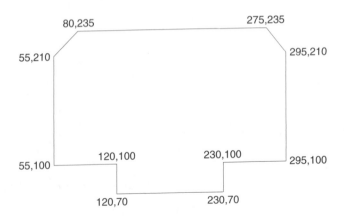

Fig. 2.6 Second example – **Line** tool

Third example – Line tool (Fig. 2.7)

1. Close the drawing and open a new **acadiso.dwt** window.
2. *Left-click* on the **Line** tool icon and *enter* figures as follows at each prompt of the command line sequence:

Command:_line Specify first point: *enter* **60,210** *right-click*
Specify next point or [Undo]: *enter* **@50,0** *right-click*
Specify next point or [Undo]: *enter* **@0,20** *right-click*
Specify next point or [Close/Undo]: *enter* **@130,0** *right-click*
Specify next point or [Close/Undo]: *enter* **@0,−20** *right-click*
Specify next point or [Close/Undo]: *enter* **@50,0** *right-click*
Specify next point or [Close/Undo]: *enter* **@0,−105** *right-click*

Specify next point or [Close/Undo]: *enter* @**−50,0** *right-click*
Specify next point or [Close/Undo]: *enter* @**0,−20** *right-click*
Specify next point or [Close/Undo]: *enter* @**−130,0** *right-click*
Specify next point or [Close/Undo]: *enter* @**0,20** *right-click*
Specify next point or [Close/Undo]: *enter* @**−50,0** *right-click*
Specify next point or [Close/Undo]: *enter* **c** (Close) *right-click*
Command:

The result is as shown in Fig. 2.7.

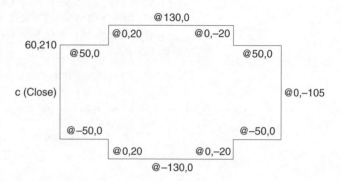

Fig. 2.7 Third example – **Line**
tool

Notes

1. The figures typed at the keyboard determining the corners of the outlines in the above examples are 2D *x,y* coordinate points. When working in 2D, coordinates are expressed in terms of two numbers separated by a comma.
2. Coordinate points can be shown as positive or as negative numbers.
3. The method of constructing an outline as shown in the first two examples is known as the **absolute coordinate entry** method, where the *x,y* coordinates of each corner of the outlines were *entered* at the command line as required.
4. The method of constructing an outline as in the third example is known as the **relative coordinate entry** method – coordinate points are *entered* relative to the previous entry. In relative coordinate entry, the @ symbol is *entered* before each set of coordinates with the following rules in mind:
   +ve *x* entry is to the right.
   −ve *x* entry is to the left.
   +ve *y* entry is upwards.
   −ve *y* entry is downwards.
5. The next example (the fourth) shows how lines at angles can be drawn taking advantage of the relative coordinate entry method. Angles in AutoCAD are measured in 360° in a counter-clockwise (anticlockwise) direction (Fig. 2.8). The < symbol precedes the angle.

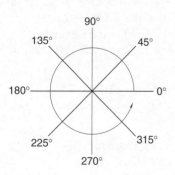

Fig. 2.8 The counter-clockwise
direction of measuring angles in
AutoCAD

Fourth example – Line tool (Fig. 2.9)

1. Close the drawing and open a new **acadiso.dwt** window.
2. *Left-click* on the **Line** tool icon and *enter* figures as follows at each prompt of the command line sequence:

> **Command:_line Specify first point:** 70,230
> **Specify next point:** @220,0
> **Specify next point:** @0,−70
> **Specify next point or [Undo]:** @115<225
> **Specify next point or [Undo]:** @−60,0
> **Specify next point or [Close/Undo]:** @115<135
> **Specify next point or [Close/Undo]:** @0,70
> **Specify next point or [Close/Undo]:** c (Close)
> **Command:**

The result is as shown in Fig. 2.9.

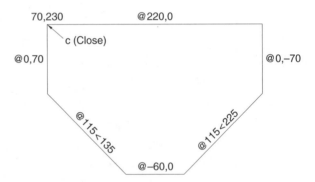

Fig. 2.9 Fourth example – **Line** tool

Fifth example – Line tool (Fig. 2.10)

Another method of constructing accurate drawings is by using a method known as **tracking**. When **Line** is in use, as each **Specify next point**: appears at the command line, a *rubber-banded* line appears from the last point *entered. Drag* the rubber-banded line in any direction and *enter* a number at the keyboard, followed by a *right-click*. The line is drawn in the *dragged* direction of a length in units equal to the *entered* number.

In this example, because all lines are drawn in either the vertical or the horizontal direction, either press the **F8** key or *click* the **ORTHO** button in the status bar.

1. Close the drawing and open a new **acadiso.dwt** window.
2. *Left-click* on the **Line** tool icon and *enter* figures as follows at each prompt of the command line sequence:

> **Command:_line Specify first point:** *enter* **65,220** *right-click*
> **Specify next point:** *drag* to right *enter* **240** *right-click*
> **Specify next point:** *drag* down *enter* **145** *right-click*
> **Specify next point or [Undo]:** *drag* left *enter* **65** *right-click*
> **Specify next point or [Undo]:** *drag* upwards *enter* **25** *right-click*

Specify next point or [Close/Undo]: *drag* left *enter* **120** *right-click*
Specify next point or [Close/Undo]: *drag* upwards *enter* **25** *right-click*
Specify next point or [Close/Undo]: *drag* left *enter* **55** *right-click*
Specify next point or [Close/Undo]: **c** (Close) *right-click*
Command:

The result is as shown in Fig. 2.10.

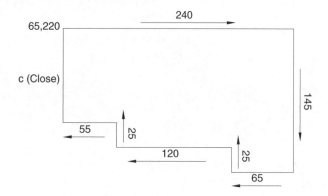

Fig. 2.10 Fifth example – **Line** tool

Fig. 2.11 The **Circle** tool icon in
the **Draw** toolbar with its tool tip

Drawing with the Circle tool

First example – Circle tool (Fig. 2.13)

1. Close the drawing just completed and open the **acadiso.dwt** screen.
2. *Left-click* on the **Circle** tool icon in the **Draw** toolbar (Fig. 2.11).
3. Type numbers against the prompts appearing in the command window as shown in Fig. 2.12, followed by *right-clicks*. The circle (Fig. 2.13) appears on screen.

Fig. 2.12 First example – **Circle**.
The command line prompts when
Circle is called

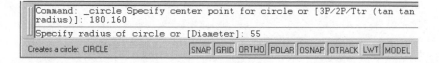

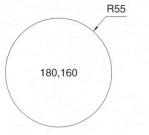

Fig. 2.13 First example – **Circle**
tool

Second example – Circle tool (Fig. 2.11)

1. Close the drawing and open the **acadiso.dwt** screen.
2. *Left-click* on the **Circle** tool icon and construct two circles as shown in the drawing Fig. 2.14.

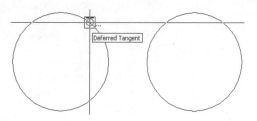

Fig. 2.14 Second example – **Circle**
tool. The two circles of radius 50

3. *Click* the **Circle** tool again and against the first prompt *enter* **t** (the abbreviation for the prompt **tan tan radius**), followed by a *right-click*.

> **Command:_circle Specify center point for circle or [3P/2P/Ttr (tan**
> **tan radius)]:** *enter* **t** *right-click*
> **Specify point on object for first tangent of circle:** *pick*
> **Specify point on object for second tangent of circle:** *pick*
> **Specify radius of circle (50):** *enter* **40** *right-click*
> **Command:**

The radius-**40** circle, tangential to the two already drawn circles, then appears (Fig. 2.15)

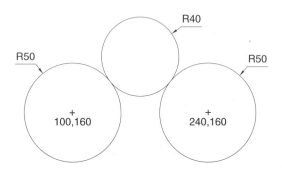

Fig. 2.15 The radius-**40** circle tangential to the radius-**50** circles

Notes

1. When a point on either circle is picked a tip appears **Deferred Tangent**. This tip will appear only when the **OSNAP** is set on with a *click* on the button, or the **F3** key of the keyboard is pressed.
2. Circles can be drawn through 3 points or through 2 points *entered* at the command line in response to prompts brought to the command line by using **3P** and **2P** in answer to the circle command line prompts.

The Erase tool

If an error has been made when using any of the **AutoCAD 2005** tools, the object or objects which have been incorrectly constructed can be deleted with the **Erase** tool. The **Erase** tool icon is at the top of the **Modify** toolbar (Fig. 2.16).

First example – Erase (Fig. 2.18)

1. With **Line** tool construct the outline in Fig. 2.17.
2. Assuming two lines of the outline have been incorrectly drawn, *left-click* the **Erase** tool icon. The command line shows:

> **Command:_erase**
> **Select objects:** *pick* one of the lines
> **Select objects:** *pick* the other line
> **Select objects:** *right-click*
> **Command:**

Fig. 2.16 The **Erase** tool icon and its tool tip at the top of the **Modify** toolbar

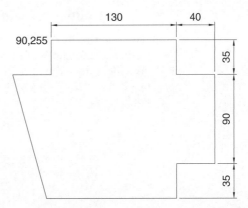

Fig. 2.17 First example – **Erase**.
An incorrect outline

And the two lines are deleted (right-hand drawing of Fig. 2.18).

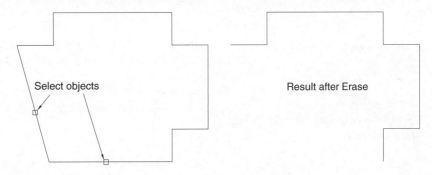

Fig. 2.18 First example – **Erase**

Second example – Erase (Fig. 2.19)

The two lines could also have been deleted by the following method:

1. *Left-click* the **Erase** tool icon. The command line shows:

> **Command:_erase**
> **Select objects:** *enter* c (Crossing)
> **Specify first corner:** *pick* **Specify opposite corner:** *pick* **2 found**
> **Select objects:** *right-click*
> **Command:**

And the two lines are deleted as in the right-hand drawing of Fig. 2.18.

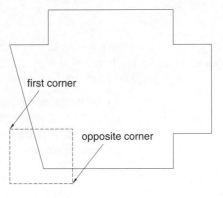

Fig. 2.19 Second example – **Erase**

Undo and Redo tools

Two other tools of value when errors have been made are the **Undo** and **Redo** tools. To undo the last action taken by any tool when constructing a drawing, either *left-click* the **Undo** tool (Fig. 2.20) or type **u** at the command line. No matter which method is adopted the error is deleted from the drawing.

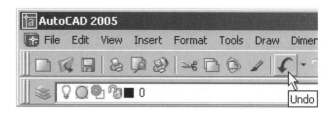

Fig. 2.20 The **Undo** tool in the **Standard** toolbar

Fig. 2.21 The **Redo** tool in the **Standard** toolbar

Everything constructed during a session in constructing a drawing can be undone by repeated *clicking* on the **Undo** tool icon or by *entering* **u**'s at the command line.

To bring back objects that have been removed by the use of **Undo**s, *left-click* the **Redo** tool icon (Fig. 2.21) or *enter* **redo** at the command line.

Drawing with the Polyline tool

Possibly the most versatile of the tools from the **Draw** toolbar. When drawing lines with the **Line** tool, each line drawn is an object in its own right, so a rectangle drawn with the **Line** tool is made up of four objects. But, a rectangle drawn with the **Polyline** tool is a single object. Lines of different thickness, arcs, arrows and circles can all be drawn using this tool as will be shown in the examples describing constructions using the **Polyline** tool. Constructions resulting from using the tool are known as **polylines** or **plines**.

To call the **Polyline** tool for use, *left-click* on its tool icon from the **Draw** toolbar (Fig. 2.9).

First example – Polyline tool (Fig. 2.23)

Note

Fig. 2.22 The **Polyline** tool from the **Draw** toolbar

In this example *enter* and *right-click* have been left out from the command line responses.

Left-click the **Polyline** tool (Fig. 2.22). The command line shows:

> **Command:_pline Specify start point:** 30,250
> **Current line width is 0**
> **Specify next point or [Arc/Halfwidth/Length/Undo/Width]:** 230,250
> **Specify next point or [Arc/Close/Halfwidth/Length/Undo/Width]:** 230,120

Specify next point or [Arc/Close/Halfwidth/Length/Undo/Width]:
30,120
Specify next point or [Arc/Close/Halfwidth/Length/Undo/Width]:
c (Close)
Command:

Notes

1. Note the prompts – **Arc** for constructing pline arcs; **Close** to close an outline; **Halfwidth** to halve the width of a wide pline; **Length** to *enter* the required length of a pline; **Undo** to undo the last pline constructed; **Close** to close an outline.
2. Only the capital letter of a prompt needs to be *entered* to make that prompt effective.
3. Other prompts will appear when the **Polyline** tool is in use as will be shown in later examples.

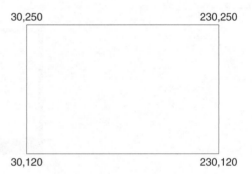

Fig. 2.23 First example –
Polyline tool

Second example – Polyline tool (Fig. 2.24)

This will be a long sequence, but it is typical of a reasonably complex drawing using the **Polyline** tool. In the following sequences, when a prompt line is to be repeated, the prompts in square brackets ([ ]) will be replaced by [**prompts**].
Left-click the **Polyline** tool icon. The command line shows:

Command:_pline Specify start point: 40,250
Current line width is 0
Specify next point or [Arc/Halfwidth/Length/Undo/Width]: w (Width)
Specify starting width <0>: 5
Specify ending width <5>: *right-click*
Specify next point or [Arc/Close/Halfwidth/Length/Undo/Width]:
160,250
Specify next point or [prompts]: h (Halfwidth)
Specify starting half-width <2.5>: 1
Specify ending half-width <1>: *right-click*
Specify next point or [prompts]: 260,250
Specify next point or [prompts]: 260,180

Specify next point or [prompts]: w (Width)
Specify starting width <1>: 10
Specify ending width <10>: *right-click*
Specify next point or [prompts]: 260,120
Specify next point or [prompts]: h (Halfwidth)
Specify starting half-width <5>: 2
Specify ending half-width <2>: *right-click*
Specify next point or [prompts]: 160,120
Specify next point or [prompts]: w (Width)
Specify starting width <4>: 20
Specify ending width <20>: *right-click*
Specify next point or [prompts]: 40,120
Specify starting width <20>: 5
Specify ending width <5>: *right-click*
Specify next point or [prompts]: cl (CLose)
Command:

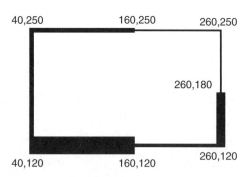

Fig. 2.24 Second example –
Polyline tool

Third example – Polyline tool (Fig. 2.25)

Left-click the **Polyline** tool icon. The command line shows:

Command:_pline Specify start point: 50,220
Current line width is 0
[prompts]: w (Width)
Specify starting width <0>: 0.5
Specify ending width <0.5>: *right-click*
Specify next point or [prompts]: 120,220
Specify next point or [prompts]: a (Arc)
Specify endpoint of arc or [prompts]: s (second pt)
Specify second point on arc: 150,200
Specify end point of arc: 180,220
Specify end point of arc or [prompts]: l (Line)
Specify next point or [prompts]: 250,220
Specify next point or [prompts]: 250,190
Specify next point or [prompts]: a (Arc)
Specify endpoint of arc or [prompts]: s (second pt)
Specify second point on arc: 240,170

Specify end point of arc: 250,150
Specify end point of arc or [prompts]: l (Line)
Specify next point or [prompts]: 250,150
Specify next point or [prompts]: 250,120
Command:

And so on until the outline Fig. 2.25 is completed.

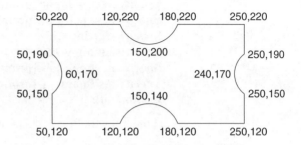

Fig. 2.25 Third example –
Polyline tool

Fourth example – Polyline tool (Fig. 2.26)

Left-click the **Polyline** tool icon. The command line shows:

Command:_pline Specify start point: 80,170
Current line width is 0
Specify next point or [prompts]: w (Width)
Specify starting width <0>: 1
Specify ending width <1>: *right-click*
Specify next point or [prompts]: a (Arc)
Specify endpoint of arc or [prompts]: s (second pt)
Specify second point on arc: 160,250
Specify end point of arc: 240,170
Specify end point of arc or [prompts]: cl (CLose)
Command:

And the circle in Fig. 2.26 is formed.

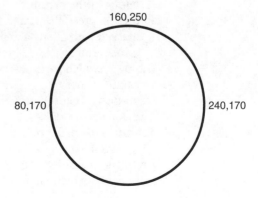

Fig. 2.26 Fourth example –
Polyline tool

Fifth example – Polyline tool (Fig. 2.27)

Left-click the **Polyline** tool icon. The command line shows:

> **Command:_pline Specify start point:** 60,180
> **Current line width is 0**
> **Specify next point or [prompts]:** w (Width)
> **Specify starting width <0>:** 1
> **Specify ending width <1>:** *right-click*
> **Specify next point or [prompts]:** 190,180
> **Specify next point or [prompts]:** w (Width)
> **Specify starting width <1>:** 20
> **Specify ending width <20>:** 0
> **Specify next point or [prompts]:** 265,180
> **Specify next point or [prompts]:** *right-click*
> **Command:**

And the arrow in Fig. 2.27 is formed.

Fig. 2.27 Fifth example –
Polyline tool

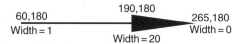

60,180   190,180   265,180
Width = 1   Width = 0
                 Width = 20

Revision notes

The following terms have been used in this chapter:

Left-click – press the left-hand button of the mouse.

Click – same meaning as *left-click*.

Double-click – press the left-hand button of the mouse twice in quick succession.

Right-click – press the right-hand button of the mouse; it has the same result as pressing the **Return** key of the keyboard.

Drag – move the cursor on to an object and, holding down the right-hand button of the mouse, pull the object to a new position.

Enter – type the letters, or numbers which follow at the keyboard.

Pick – move the cursor on to an item on screen and press the left-hand button of the mouse.

Return – press the *Enter* key of the keyboard. This key may also be marked with a left-facing arrow. In most cases (but not always) has the same result as a *right-click*.

Dialog – a window appearing in the AutoCAD window in which settings can be made.

Drop-down menu – a menu appearing when one of the names in the menu bars is *clicked*.

Tooltip – the name of a tool appearing when the cursor is placed over a tool icon from a toolbar.

Prompts – text appearing in the command window when a tool is selected, which advise the operator as to which operation is required.

Methods of coordinate entry – Three methods of coordinate entry have been used in this chapter:

1. **Absolute method** – the coordinates of points on an outline are *entered* at the command line in response to prompts.
2. **Relative method** – the distances in coordinate units are *entered* preceded by @ from the last point which has been determined on an outline. Angles which are measured in a counter-clockwise direction are preceded by >.
3. **Tracking** – the rubber band of the tool is *dragged* in the direction in which the line is to be drawn and its distance in units is *entered* at the command line followed by a *right-click*.

Line and Polyline tools – an outline drawn using the **Line** tool consists of a number of objects equal to the number of lines in the outline. An outline drawn using the **Polyline** is a single object no matter how many plines are in the outline.

Exercises

1. Using the **Line** tool construct the rectangle in Fig. 2.28.

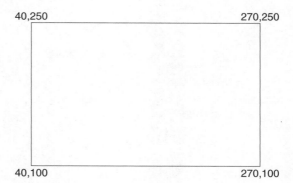

Fig. 2.28 Exercise 1

2. Construct the outline in Fig. 2.29 using the **Line** tool. The coordinate points of each corner of the rectangle will need to be calculated from the lengths of the lines between the corners.

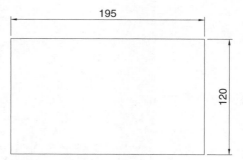

Fig. 2.29 Exercise 2

3. Using the **Line** tool, construct the outline in Fig. 2.30.

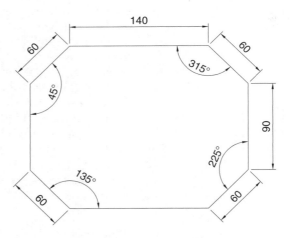

Fig. 2.30 Exercise 3

4. Using the **Circle** tool, construct the two circles of radius 50 and 30. Then using the **Ttr** prompt add the circle of radius 25 (Fig. 2.31).

Fig. 2.31 Exercise 4

5. Fig. 2.32. In an **acadiso.dwt** screen, using the **Circle** and **Line** tools, construct the line and the circle of radius 40. Then, using the **Ttr** prompt, add the circle of radius 25.

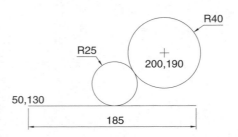

Fig. 2.32 Exercise 5

6. Using the **Line** tool, construct the two lines at the length and angle as given in Fig. 2.33. Then with the **Ttr** prompt of the **Circle** tool, add the circle as shown.

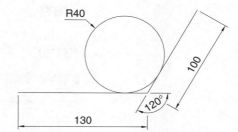

Fig. 2.33 Exercise 6

7. Using the **Polyline** tool, construct the outline given in Fig. 2.34.

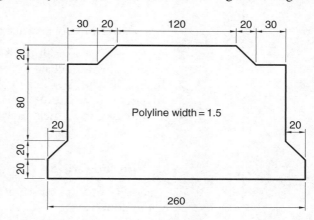

Fig. 2.34 Exercise 7

8. Construct the outline in Fig. 2.35 using the **Polyline** tool.

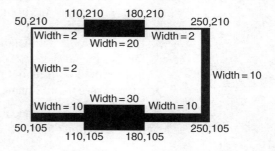

Fig. 2.35 Exercise 8

9. With the **Polyline** tool construct the arrows shown in Fig. 2.36.

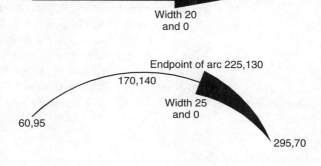

Fig. 2.36 Exercise 9

CHAPTER 3

Osnap, AutoSnap and Draw tools

Fig. 3.1 The tool names in the **Draw** drop-down menu

Fig. 3.2 The **Arc** tool icon in the **Draw** toolbar

Aims of this chapter

1. To describe the uses of the **Arc**, **Ellipse**, **Polygon** and **Rectangle** tools from the **Draw** toolbar.
2. To describe the uses of the **Polyline Edit** (pedit) tool.
3. To introduce the **AutoSnap** system and its uses.
4. To introduce the **Object Snap** (osnap) system and it uses.

Introduction

The majority of tools in AutoCAD 2005 can be called into use in the following four ways:

1. With a *click* on the tool's icon in its toolbar.
2. By *clicking* on the tool's name in an appropriate drop-down menu. Fig. 3.1 shows the tool names displayed in the **Draw** drop-down menu.
3. By *entering* an abbreviation for the tool name at the command line in the Command palette. For example the abbreviation for the **Line** tool is **l**, for the **Polyline** tool it is **pl** and for the **Circle** tool it is **c**.
4. By *entering* the full name of the tool at the command line.

In practice operators constructing drawings in AutoCAD 2005 may well use a combination of these four methods.

The Arc tool

In AutoCAD 2005, arcs can be constructed using any three of the following characteristics of an arc: Its **Start** point; a point on the arc (**Second** point); its **Center**; its **End**; its **Radius**; **Length** of the arc; **Direction** in which the arc is to be constructed; **Angle** between lines of the arc.

In the examples which follow, *entering* initials for these characteristics, in response to prompts at the command line when the **Arc** tool is called, allows arcs to be constructed in a variety of ways.

To call the **Arc** tool *click* on its tool icon in the **Draw** toolbar (Fig. 3.2), or *click* on **Arc** in the **Draw** drop-down menu. A sub-menu shows the

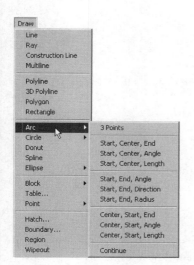

Fig. 3.3 The **Arc** sub-menu of the **Draw** drop-down menu

possible methods of constructing arcs (Fig. 3.3). The abbreviation for calling the **Arc** tool is **a**.

First example – Arc tool (Fig. 3.4)

Left-click the **Arc** tool icon. The command line shows:

Command:_arc Specify start point of arc or [Center]: 100,220
Specify second point of arc or [Center/End]: 55,250
Specify end point of arc: 10,220
Command:

Second example – Arc tool (Fig. 3.4)

Command: *right-click* brings back the **Arc** sequence
ARC Specify start point of arc or [Center]: c (Center)
Specify center point of arc: 200,190
Specify start point of arc: 260,215
Specify end point of arc or [Angle/chord Length]: 140,215
Command:

Third example – Arc tool (Fig. 3.4)

Command: *right-click* brings back the **Arc** sequence
ARC Specify start point of arc or [Center]: 420,210
Specify second point of arc or [Center/End]: e (End)
Specify end point of arc: 320,210
Specify center point of arc or [Angle/Direction/Radius]: r (Radius)
Specify radius of arc: 75
Command:

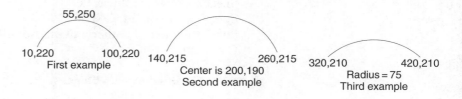

Fig. 3.4 Examples – **Arc** tool

The Ellipse tool

Ellipses can be regarded as what is seen when a circle is viewed from directly in front of the circle and when the circle is rotated through an angle about its horizontal diameter. Ellipses are measured in terms of two axes – a **major axis** and a **minor axis**; the major axis being the diameter of the circle, the minor axis being the height of the ellipse after the circle has been rotated through an angle (Fig. 3.5).

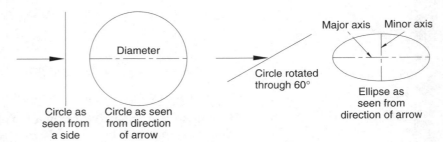

Fig. 3.5 An ellipse can be regarded as viewing a rotated circle

Fig. 3.6 The **Ellipse** tool icon in the **Draw** toolbar

To call the **Ellipse** tool, *click* on its tool icon in the **Draw** toolbar (Fig. 3.6) or *click* its name in the **Draw** drop-down menu. The abbreviation for calling the **Ellipse** tool is **el**.

First example – Ellipse (Fig. 3.7)

Left-click the **Ellipse** tool icon. The command line shows:

Command:_ellipse
Specify axis endpoint of elliptical arc or [Center]: 30,190
Specify other endpoint of axis: 150,190
Specify distance to other axis or [Rotation]: 25
Command:

Second example – Ellipse (Fig. 3.7)

In this second example, the coordinates of the centre of the ellipse (the point where the two axes intersect) are *entered*, followed by *entering* coordinates for the end of the major axis, followed by *entering* the units for the end of the minor axis.

Command: *right-click*
ELLIPSE
Specify axis endpoint of elliptical arc or [Center]: c (Center)
Specify center of ellipse: 260,190
Specify endpoint of axis: 205,190
Specify distance to other axis or [Rotation]: 30
Command:

Third example – Ellipse (Fig. 3.7)

In this third example, after setting the positions of the ends of the major axis, the angle of rotation of the circle from which an ellipse can be obtained is *entered*.

Command: *right-click*
ELLIPSE
Specify axis endpoint of elliptical arc or [Center]: 30,100
Specify other endpoint of axis: 120,100
Specify distance to other axis or [Rotation]: r (Rotation)
Specify rotation around major axis: 45
Command:

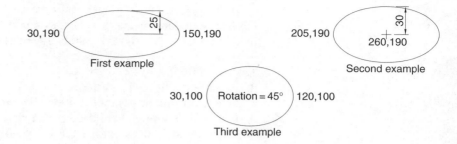

Fig. 3.7 Examples – **Ellipse**

Saving drawings

Before going further it is as well to know how to save the drawings constructed when answering examples and exercises in this book. When a drawing has been constructed, *left-click* on **File** in the menu bar and on **Save As** . . . in the drop-down menu (Fig. 3.8). The **Save Drawing As** dialog appears (Fig. 3.9).

Fig. 3.8 Selecting **Save As** . . . in the **File** drop-down menu

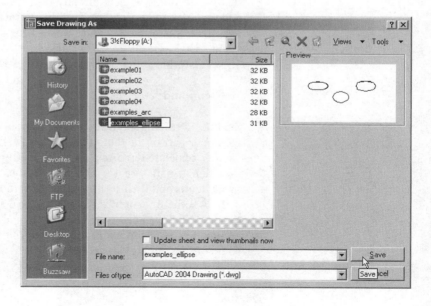

Fig. 3.9 The **Save Drawing As** dialog

Unless you are the only person to use the computer on which the drawing has been constructed, it is best to save work to a floppy disk, usually held in the drive **A:**. To save a drawing to a floppy in drive **A**:

1. Place a floppy disk in drive **A:**.
2. In the **Save in:** field of the dialog, *click* the arrow to the right of the field and from the popup list select **3½ Floppy [A:]**.
3. In the **File name:** field of the dialog, type a suitable name. The file name extension **.dwg** does not need to be typed because it will automatically be added to the file name.
4. *Left-click* the **Save** button of the dialog. The drawing will be saved to the floppy with the file name extension **.dwg** – the AutoCAD file name extension.

Using Snap

In previous chapters several methods of constructing accurate drawings have been described – using **Snap**; absolute coordinate entry; relative coordinate entry and tracking.

Other methods of ensuring accuracy between parts of constructions are by making use of **Object Snaps** (**Osnaps**) and **AutoSnap**.

Snap, **Grid** and **Osnap** can be set from the buttons in the status bar or by keying **F3** (**Osnap**), **F7** (**Grid**) and **F9** (**Snap**).

Object Snaps (Osnaps)

Osnaps allow objects to be added to a drawing at precise positions in relation to other objects already on screen. With osnaps, objects can be added to the end points, mid points, to intersections of objects, to centres and quadrants of circles and so on. Osnaps also override snap points even when snap is set on.

To set **Osnaps**, at the command line:

Command: *enter* **os**

And the **Drafting Settings** dialog appears. *Click* the **Object Snap** tab in the upper part of the dialog and *click* in each of the check boxes (the small squares opposite the osnap names). See Fig. 3.10.

When osnaps are set **ON**, as outlines are constructed using osnaps, osnap icons and their tooltips appear as indicated in Fig. 3.11.

It is sometimes advisable not to have **Osnaps** set on in the **Drafting Settings** dialog, but to set **Osnap** off and use osnap abbreviations at the command line when using tools. The following examples show the use of some of these abbreviations.

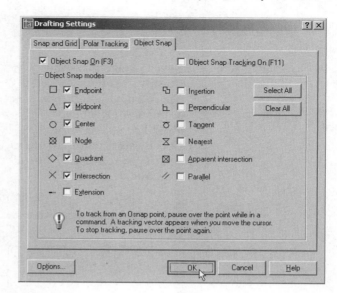

Fig. 3.10 The **Drafting Settings** dialog with some **Osnaps** set on

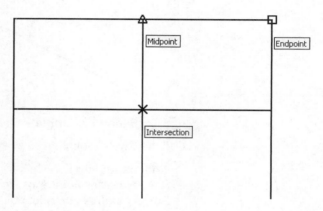

Fig. 3.11 Three osnap icons and their tooltips

First example – Osnap abbreviations (Fig. 3.12)

Call the **Polyline** tool:

> **Command:_pline**
> **Specify start point:** 50,230
> **[prompts]:** w (Width)
> **Specify starting width:** 1
> **Specify ending width <1>:** *right-click*
> **Specify next point:** 260,230
> **Specify next point:** *right-click*
> **Command:** *right-click*
> **PLINE**
> **Specify start point:** end **of** *pick* the right-hand end of the pline
> **Specify next point:** 50,120

Specify next point: *right-click*
Command: *right-click*
PLINE
Specify start point: mid **of** *pick* near the middle of the first pline
Specify next point: 155,120
Specify next point: *right-click*
Command: *right-click*
PLINE
Specify start point: int **of** *pick* the plines at their intersection
Specify next point: *right-click*
Command:

The result is as shown in Fig. 3.12. In this illustration the osnap tooltips are shown as they appear when each object is added to the outline.

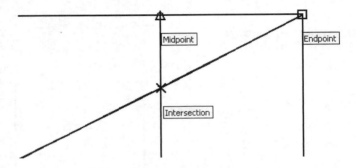

Fig. 3.12 First example – **Osnaps**

Second example – Osnap abbreviations (Fig. 3.13)

Call the **Circle** tool:

Command:_circle
Specify center point for circle: 180,170
Specify radius of circle: 60
Command: *enter* **l** (Line) *right-click*
Specify first point: *enter* **qua** *right-click*
of *pick* near the upper quadrant of the circle
Specify next point: *enter* **cen** *right-click*

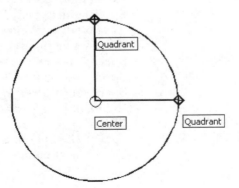

Fig. 3.13 Second example –
Osnaps

of *pick* near the centre of the circle
Specify next point: *enter* **qua** *right-click*
of *pick* near right-hand side of circle
Specify next point: *right-click*
Command:

Note

With osnaps off, the following abbreviations can be used:

end   endpoint;
int   intersection;
qua   quadrant;
ext   extension;
mid   midpoint;
cen   centre;
nea   nearest.

AutoSnap

AutoSnap is similar to **Osnap**. To set **AutoSnap**, *right-click* in the command window and from the menu which appears *click* **Options** . . . The **Options** dialog appears. *Click* the **Drafting** tab in the upper part of the dialog and set the check boxes against the **AutoSnap Settings** on (tick in boxes). These settings are shown in Fig. 3.14.

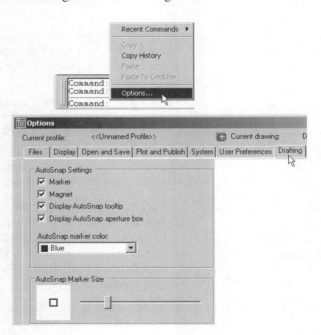

Fig. 3.14 Setting **AutoSnap** in the **Options** dialog

With **AutoSnap** set, each time an object is added to a drawing the **AutoSnap** features appear as indicated in Fig. 3.15.

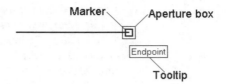

Fig. 3.15 The features of **AutoSnap**

Part of a drawing showing the features of a number of **AutoSnap** points is given in Fig. 3.16.

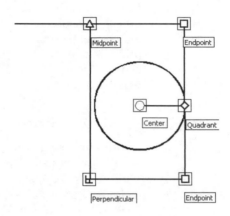

Fig. 3.16 A number of **AutoSnap** features

Note

OSNAP must be set **ON** for the AutoSnap features to show, when constructing a drawing with their aid.

Examples of using some Draw tools

First example – Polygon tool (Fig. 3.18)

1. Call the **Polygon** tool – either with a *click* on its tool icon in the **Draw** toolbar (Fig. 3.17), or by *entering* **pol** or **polygon** at the command line, or from the **Draw** drop-down menu. The command line shows:

Fig. 3.17 The **Polygon** tool icon in the **Draw** toolbar

Command:_polygon Enter number of sides <4>: 6
Specify center of polygon or [Edge]: 60,210
Enter an option [Inscribed in circle/Circumscribed about circle]
    <I>: *right-click* (accept Inscribed)
Specify radius of circle: 60
Command:

2. In the same manner construct a **5**-sided polygon of centre **200,210** and radius **60**.
3. Then, construct an **8**-sided polygon of centre **330,210** and radius **60**.
4. Repeat to construct a **9**-sided polygon circumscribed about a circle of radius **60** and centre **60,80**.

5. Construct yet another polygon with **10** sides of radius **60** and centre **200,80**.

6. Finally another polygon circumscribing a circle of radius **60**, centre **330,80** and sides **12**.

The result is shown in Fig. 3.18.

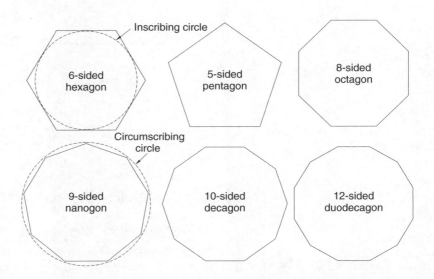

Fig. 3.18 First example –
Polygon tool

Fig. 3.19 The **Rectangle** tool in
the **Draw** toolbar

Second example – Rectangle tool (Fig. 3.20)

Call the **Rectangle** tool – either with a *click* on its tool icon in the **Draw** toolbar (Fig. 3.19), or by *entering* **rec** or **rectang** at the command line, or from the **Draw** drop-down menu. The command line shows:

Command:_rectang
Specify first corner point or [Chamfer/Elevation/Fillet/Thickness/ Width]: 25,240
Specify other corner point or [Dimensions]: 160,160
Command:

Third example – Rectangle tool (Fig. 3.20)

Command:_rectang
[prompts]: c (Chamfer)
Specify first chamfer distance for rectangles <0>: 15
Specify first chamfer distance for rectangles <15>: *right-click*
Specify first corner point: 200,240
Specify other corner point: 300,160
Command:

Fourth example – Rectangle (Fig. 3.20)

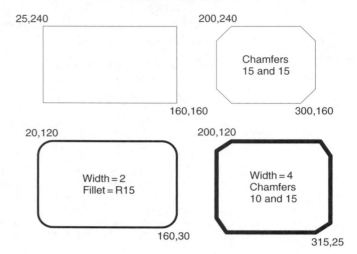

Fig. 3.20 Examples – **Rectangle**
tool

The Polyline Edit tool

Polyline Edit or **Pedit** is a valuable tool for editing plines.

First example – Polyline Edit (Figs 3.23 and 3.24)

1. With the **Polyline** tool construct the outlines **1** to **6** of Fig. 3.23.
2. Call the **Edit Polyline** tool – either with a *click* on its tool icon in the **Modify II** toolbar (Fig. 3.21), or by *entering* **pe** or **pedit** at the command line, or from the **Modify** drop-down menu (Fig. 3.22).

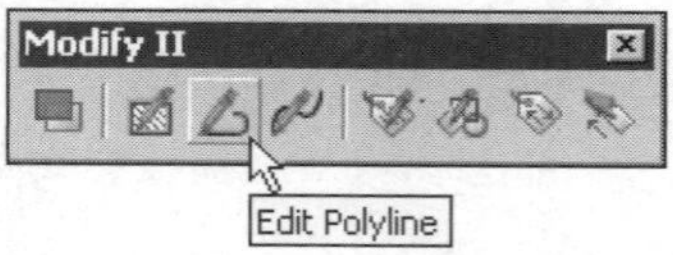

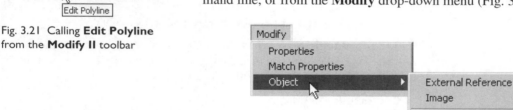

Fig. 3.21 Calling **Edit Polyline**
from the **Modify II** toolbar

Fig. 3.22 Calling **Polyline Edit**
from the **Modify** drop-down menu

By far the easiest method of calling the tool is to *enter* **pe** at the command line. The command line shows:

Command: *enter* pe
PEDIT Select polyline or [Multiple]: *pick* pline **2**
**Enter an option [Open/Join/Width/Edit vertex/Fit/Spline/Decurve/
    Ltype gen/Undo]:** w (Width)
Specify new width for all segments: 2
**Enter an option [Open/Join/Width/Edit vertex/Fit/Spline/Decurve/
    Ltype gen/Undo]:** *right-click*
Command:

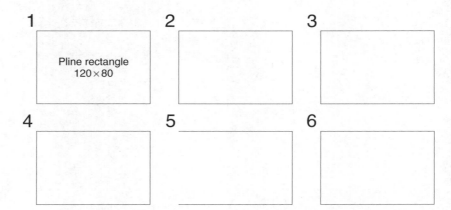

Fig. 3.23 Example – **Polyline Edit**

3. Repeat with pline **3** and pedit to Width = **10**.
4. Repeat with pline **4** and *enter* **s** (Spline) in response to the prompt line:

Enter an option [Open/Join/Width/Edit vertex/Fit/Spline/Decurve/ Ltype gen/Undo]:

5. Repeat with pline **5** and *enter* **j** (Join) in response to the prompt line:

Enter an option [Open/Join/Width/Edit vertex/Fit/Spline/Decurve/ Ltype gen/Undo]:

The result is shown in pline **6**.

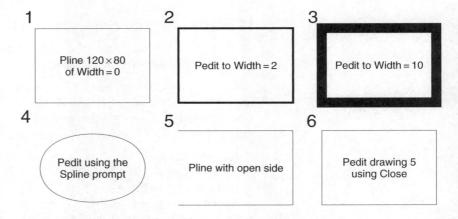

Fig. 3.24 Example – **Polyline Edit**

Example – Multiple Polyline Edit (Fig. 3.25)

1. With the **Polyline** tool construct the left-hand outlines of Fig. 3.25.
2. Call the **Edit Polyline** tool. The command line shows:

Command: *enter* pe
PEDIT Select polyline or [Multiple]: m (Multiple)
Select objects: *pick* any one of the lines or arcs of Fig. 3.25
    1 found
Select objects: *pick* another line or arc **1 found 2 total**

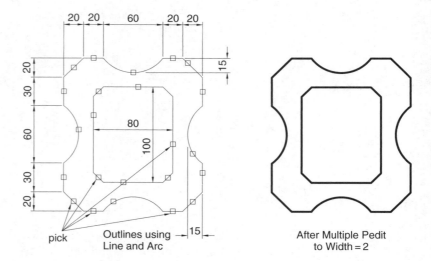

Fig. 3.25 Example – **Multiple Polyline Edit**

Continue selecting lines and arcs as shown by the *pick* boxes of the left-hand drawing of Fig. 3.25 until the command line shows:

Select objects: *pick* another line or arc **1 found 24 total**
Select objects: *right-click*
[prompts]: w (Width)
Specify new width for all segments: 1.5
[prompts]: *right-click*
Command:

The result is shown in the right-hand drawing of Fig. 3.25.

Transparent commands

When any tool is in operation it can be interrupted by prefixing the interrupting command with an apostrophe (') . This is particularly useful when wishing to zoom when constructing a drawing (see page 48). As an example when the **Line** tool is being used:

Command:_line
Specify first point: 100,120
Specify next point: 190,120
Specify next point: *enter* 'z (Zoom)
>> Specify corner of window or [prompts]: *pick*
>>>>Specify opposite corner: *pick*
Resuming line command.
Specify next point:

And so on. The transparent command method can be used with any tool.

The set variable PELLIPSE

Many of the operations performed in AutoCAD are carried out under the settings of **set variables**. Some of the numerous set variables available in

AutoCAD 2005 will be described in later pages. The variable **PELLIPSE** controls whether ellipses are drawn as splines or as polylines. It is set as follows:

> **Command:** *enter* **pellipse** *right-click*
> **Enter new value for PELLIPSE <0>:** *enter* **1** *right-click*
> **Command:**

And now when ellipses are drawn they are plines. If the variable is set to **0**, the ellipses will be splines.

Revision notes

The following terms have been used in this chapter:

Field – a part of a window or dialog in which numbers or letters are *entered* or can be read.

Popup list – a list brought in screen with a *click* on the arrow often found at the right-hand end of a field.

Object – a part of a drawing which can be treated as a single object. For example a line constructed with the **Line** tool is an object; a rectangle constructed with the **Polyline** tool is an object; an arc constructed with the **Arc** tool is an object. It will be seen in a later chapter (Chapter 9) that several objects can be formed into a single object.

Toolbar – a collection of tool icons, all of which have similar functions. For example the **Draw** toolbar contains tool icons of tools which are used for drawing. The **Modify** toolbar contains tool icons of tools used for modifying parts of drawings.

Command line – a line in the command window which commences with the word **Command:**.

Snap, **Grid** and **Osnap** can be toggled with *clicks* on the buttons in the status bar. These functions can also be set with function keys: **Snap** – **F9**; **Grid** – **F7** and **Osnap** – **F3**.

Osnaps ensure accurate positioning of objects in drawings.

AutoSnap can also be used for ensuring accurate positioning of objects in relation to other objects in a drawing.

Osnap must be set **ON** before **AutoSnap** can be used.

Osnap abbreviations can be used rather than be set from the **Drafting Settings** dialog.

Notes on tools

1. Polygons constructed with the **Polygon** tool are regular polygons – the edges of the polygons all are the same length and the angles are of the same degrees.
2. Polygons constructed with the **Polygon** tool are plines, so can be acted upon with the **Edit Polyline** tool.
3. Rectangles constructed with the **Rectangle** tool are plines. They can be drawn with chamfers or fillets and their width can be varied.

4. The **Edit Polyline** tool can be used to change plines.
5. The easiest method of calling the **Edit Polyline** tool is to *enter* **pe** at the command line.
6. The **Multiple** prompt of the **pedit** tool saves considerable time when editing a number of objects in a drawing.
7. Transparent commands can be used to interrupt tools in operation by preceding the interrupting tool name with an apostrophe (').
8. Ellipses drawn when the variable **PELLIPSE** is set to **0** are splines; when **PELLIPSE** is set to **1**, ellipses are polylines.

Exercises

1. Using the **Line** and **Arc** tools, construct the outline given in Fig. 3.26.

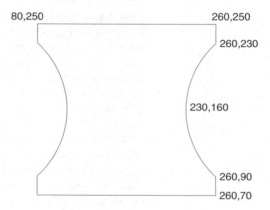

Fig. 3.26 Exercise 1

2. With the **Line** and **Arc** tools, construct the outline given in Fig. 3.27.

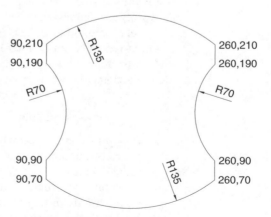

Fig. 3.27 Exercise 2

3. Using the **Ellipse** and **Arc** tools construct the drawing in Fig. 3.28.

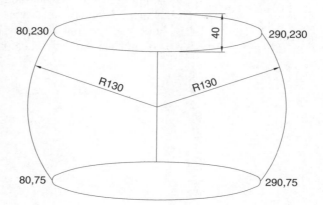

Fig. 3.28 Exercise 3

4. With the **Line**, **Circle** and **Ellipse** tools construct Fig. 3.29.

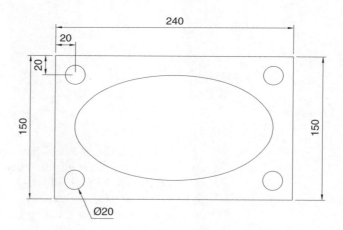

Fig. 3.29 Exercise 4

5. With the **Ellipse** tool, construct the drawing in Fig. 3.30

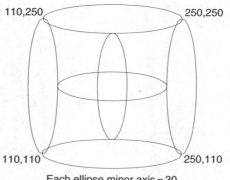

Fig. 3.30 Exercise 5

Each ellipse minor axis = 30

6. Fig. 3.31 shows a rectangle in the form of a square with hexagons along each edge. Using the **Dimensions** prompt of the **Rectangle** tool construct the square. Then, using the **Edge** prompt of the **Polygon** tool, add the four hexagons. Use the **Osnap endpoint** to ensure the polygons are in their exact positions.

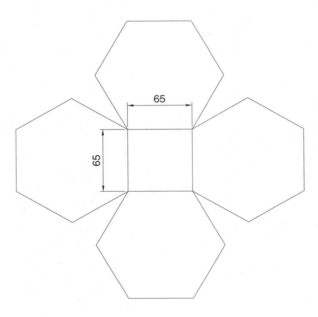

Fig. 3.31 Exercise 6

7. Fig. 3.32 shows seven hexagons with edges touching. Construct the inner hexagon using the **Polygon** tool, then with the aid of the **Edge** prompt of the tool, add the other six hexagons.

8. Fig. 3.33 was constructed using only the **Rectangle** tool. Make an exact copy of the drawing using only the **Rectangle** tool.

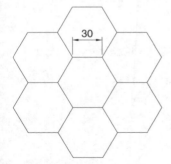

Fig. 3.32 Exercise 7

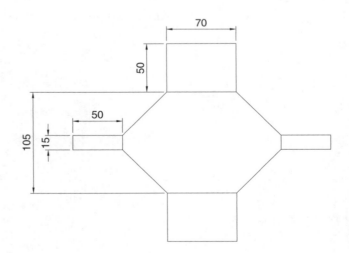

Fig. 3.33 Exercise 8

9. Construct the drawing in Fig. 3.34 using the **Line** and **Arc** tools. Then, with the aid of the **Multiple** prompt of the **Edit Polyline** tool change the outlines into plines of **Width = 1**.

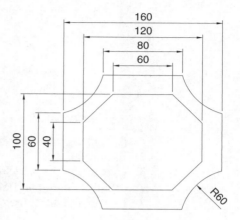

Fig. 3.34 Exercise 9

10. Construct Fig. 3.35 using the **Line** and **Arc** tools. Then change all widths of lines and arcs to a width of **2** with **Polyline Edit**.

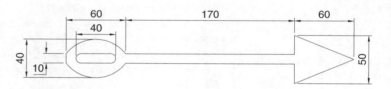

Fig. 3.35 Exercise 10

11. Construct the two outlines in Fig. 3.36 using the **Rectangle** and **Line** tools and then with **Edit Polyline** change parts of the drawing to plines of widths as shown in Fig. 3.36.

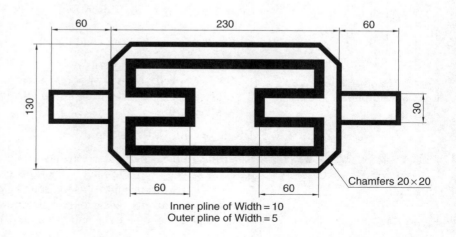

Inner pline of Width = 10
Outer pline of Width = 5

Fig. 3.36 Exercise 11

CHAPTER 4

Zoom, Pan and templates

Aims of this chapter

1. To demonstrate the value of the **Zoom** tools.
2. To introduce the **Pan** tool.
3. To describe the value of using the **Aerial View** window in conjunction with the **Zoom** and **Pan** tools.
4. To describe the construction and saving of drawing templates.

Introduction

Fig. 4.1 Calling **Zoom Realtime** from the **Standard** toolbar

The use of the **Zoom** tools not only allows the close inspection of the most minute areas of a drawing in the AutoCAD 2005 drawing area, but allows the construction of very accurate drawing of small details in a drawing.

There are several methods for calling any of the **Zoom** tools. Fig. 4.1 shows the **Zoom Realtime** tool selected from its tool icon in the **Standard** toolbar.

Fig. 4.2 shows the other **Zoom** tools held in the **Standard** toolbar. Fig. 4.3 shows all the **Zoom** tools held in the **Zoom** toolbar. Fig. 4.4 shows the flyout from the **Zoom Window** icon.

Fig. 4.3 shows the **Zoom** toolbar as selected from the menu appearing with a *right-click* in any toolbar.

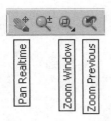

Fig. 4.2 The other **Zoom** tool icons in the **Standard** toolbar

Although any of the **Zoom** tools can be selected with a *click* on their tool icons, by far the easiest and quickest method of calling a **Zoom** tool is by *entering* **z** at the command line:

Command: *enter* **z** *right-click*
ZOOM Specify corner of window, enter a scale factor (nX or nXP)
 or [All/Center/Dynamic/Extents/Previous/Scale/Window/Object]
 <real time>:

This allows the different zooms:

Realtime – selects parts of a drawing within a window.
All – the screen reverts to the limits of the template.
Center – the drawing centres itself around a *picked* point.
Dynamic – a broken line surrounds the drawing which can be changed in size and repositioned to a part of the drawing.

Fig. 4.3 The **Zoom** toolbar

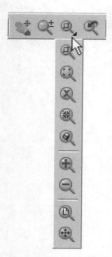

Fig. 4.4 The **Zoom Window** tool icon carries a flyout showing more **zooms**

Extents – the drawing fills the AutoCAD drawing area.
Previous – the screen reverts to its previous zoom.
Scale – entering a number or a decimal fraction scales the drawing.
Window – the parts of the drawing within a *picked* window appears on screen. The effect is the same as using **Realtime**.
Object – *pick* any object on screen and the object zooms.

The operator will probably be using **Realtime**, **Window** and **Previous** zooms most frequently.

The following illustrations show: Fig. 4.5 – drawing which has been constructed using **Polyline**, Fig. 4.6 – a **Window** zoom of part of the drawing allowing it to be checked for accuracy, Fig. 4.7 – an **Extents** zoom.

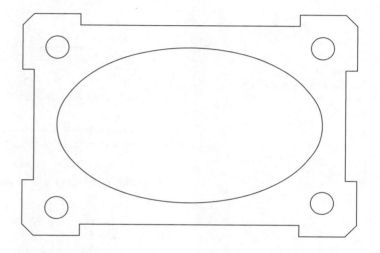

Fig. 4.5 A drawing constructed using **Polyline**

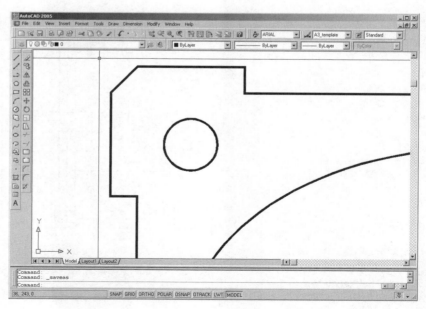

Fig. 4.6 A **Zoom Widnow** of part of the drawing in Fig. 4.5

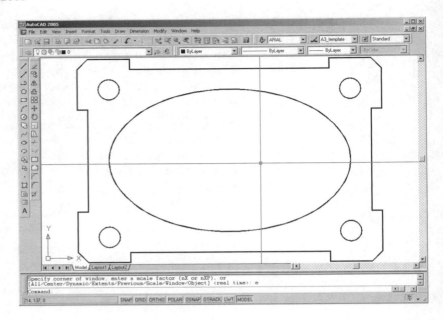

Fig. 4.7 A **Zoom Extents** of the drawing in Fig. 4.5

It will be found that the **Zoom** tools are among those most frequently used when working in AutoCAD 2005.

The Aerial View window

Left-click on **Aerial View** from the **View** drop-down menu and the **Aerial View** window appears – usually at the bottom right-hand corner of the AutoCAD 2005 window (Fig. 4.8). The **Aerial View** window shows the

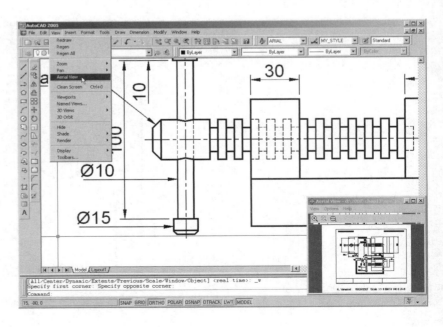

Fig. 4.8 The **Aerial View** window

whole of a drawing with that part which is within the **Limits** of the drawing template in use bounded with a thick black line. The **Aerial View** window is of value when dealing with large drawings because it allows that part of the window on screen to be shown in relation to other parts of the drawing. Fig. 4.8 shows a dimensioned three-view orthographic projection of a small bench vice.

The area of the drawing within a **Zoom** window in the drawing area is bounded by a thick black line in the **Aerial View** window.

The Pan tool

One of the tools from the **Standard** toolbar is the **Pan** tool. A *click* on its icon or *entering* **p** at the command line brings the tool into operation. When the tool is called, the cursor on screen changes to an icon of a hand. *Dragging* the hand across screen under mouse movement allows various parts of the drawing not on screen to be viewed. As the *dragging* takes place, the black rectangle in the **Aerial View** window moves in sympathy (Fig. 4.9). The **Pan** tool allows any part of the drawing to be viewed and/or modified. When that part of the drawing which is required is on screen a *right-click* calls up the menu as shown in Fig. 4.9, from which either the tool can be exited or other tools can be called.

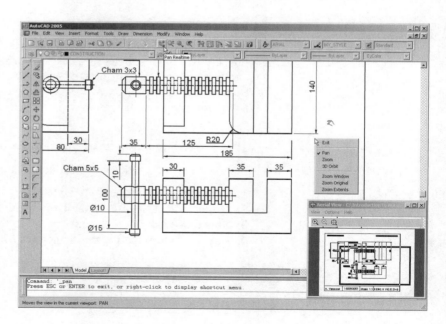

Fig. 4.9 The **Pan** tool in action showing a part of the drawing, while the whole drawing is shown in the **Aerial View** window

Notes

1. If using a mouse with a wheel, both zooms and pans can be performed with the aid of the wheel. See page 3.
2. The **Zoom** tools are important in that they allow even the smallest parts of drawings to be examined and, if necessary, amended or modified.

3. The zoom tools can be called from tool icons in the **Standard** toolbar; by selection from tool icons in the **Zoom** toolbar or by *entering* **zoom** at the command line. But easiest of all is to *enter* **z** at the command line followed by a *right-click*.
4. Similarly the easiest method of calling the **Pan** tool is to *enter* **p** at the command line followed by a *right-click*.
5. When constructing large drawings, the **Pan** tool and the **Aerial View** window are of value for allowing work to be carried out in any part of a drawing, while showing the whole drawing in the **Aerial View** window.

Drawing templates

In Chapters 1 to 3, drawings were constructed in the **acadiso.dwt** template which loaded when AutoCAD 2005 was opened. The default **acadiso** template was amended with **Limits** set to **420,297** (coordinates within which a drawing can be constructed), **Grid** set to **10**, **Snap** set to **5** and the drawing area **Zoomed** to **All**.

Throughout this book most drawings will be based on an **A3** sheet which measures 420 units by 297 units (the same as the **Limits**).

Note

As mentioned on page 11 if others are using the computer on which drawings are being constructed, it is as well to save the template being used to another file name or, if thought necessary, to a floppy disk. A file name such as **My_template.dwt**, as suggested earlier, or a name such as **book_template** can be given.

Adding features to the template

Four other features will now be added to our template:

1. **Text style** – set in the **Text Style** dialog.
2. **Dimension style** – set in the **Dimension Style Manager** dialog.
3. **Shortcut variable** – set to **0**.
4. **Layers** – set in the **Layer Properties Manager** dialog.

Setting Text

1. At the command line:

   Command: *enter* **st** (Style) *right-click*

2. The **Text Style** dialog appears (Fig. 4.10). In the dialog, *enter* **6** in the **Height** field. Then *left-click* on **Arial** in the **Font Name** popup list. **Arial** font letters appear in the **Preview** area of the dialog.
3. *Left-click* the **New** button and *enter* **ARIAL** in the **New Text Style** sub-dialog which appears (Fig. 4.11) and *click* the **OK** button.

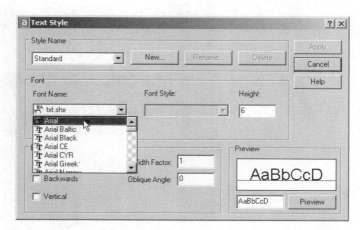

Fig. 4.10 The **Text Style** dialog

4. *Left-click* the **Close** button of the **New Text Style** dialog.

5. *Left-click* the **OK** button of the dialog.

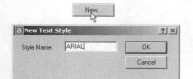

Fig. 4.11 The **New Text Style** sub-dialog

Setting Dimension style

Settings for dimensions require making *entries* in a number of sub-dialogs in the **Dimension Style Manager**. To set the dimensions style:

1. At the command line:

Command: *enter* **d** *right-click*

And the **Dimensions Style Manager** dialog appears (Fig. 4.12).

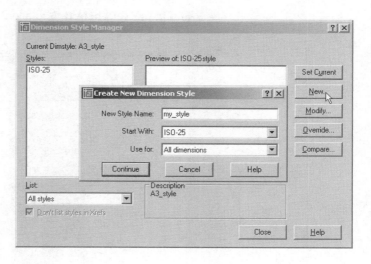

Fig. 4.12 The **Dimensions Style Manager** dialog

2. In the dialog, *click* the **New** . . . button. In the **Create New Dimension Style** sub-dialog which appears, *enter* **my_style** in the **New Style Name** field, followed by a *click* on the sub-dialog's **Continue** button.

3. The **New Dimension Style** sub-dialog appears (Fig. 4.13). In the dialog make settings as shown. Then *click* the **OK** button of that dialog.

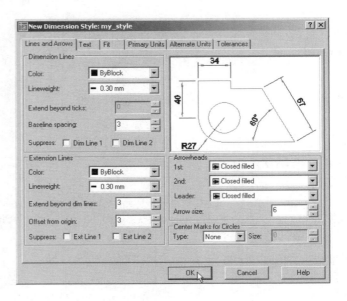

Fig. 4.13 The **New Dimension Style** sub-dialog

4. The original **Dimension Style Manager** reappears. *Click* its **Modify** button.
5. The **Modify Dimension Style** sub-dialog appears (Fig. 4.14). *Click* the **Text** tab at the top of the dialog. Then *click* the arrow to the right of the **Text Style** field and select **ARIAL** from the popup list. *Enter* a height of **6** in the **Text height** field and **2** in the **Offset from dim line** field.

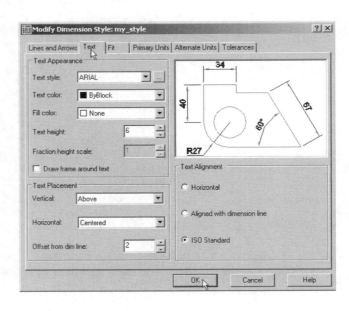

Fig. 4.14 Setting text style and height in the **Text** sub-dialog

6. Then *click* the **Primary Units** tab and set the units' **Precision** in both, **Linear** and **Angular Dimensions** to **0**, that is, no units after decimal point. *Click* the sub-dialog's **OK** button (Fig. 4.15). The **Dimension Style Manager**, dialog reappears showing dimensions, as they will appear in a drawing, in the **Preview** of **my_style** box. *Click* the **Set Current** button, followed by another *click* on the **Close** button.

Fig. 4.15 Setting units in the **Primary Units** sub-dialog

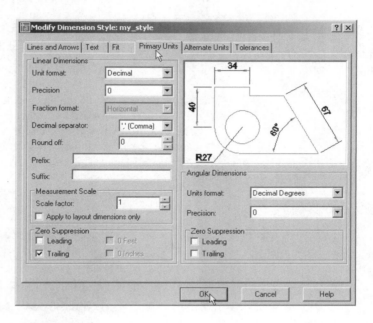

The SHORTCUTMENU variable

Call the line tool, draw a few lines and then *right-click*. The *right-click* menu shown in Fig. 4.16 may well appear. The menu may also appear when any tool is called. Some operators prefer using this menu when constructing drawings. To stop this menu from appearing:

> **Command:** *enter* **shortcutmenu** *right-click*
> **Enter new value for SHORTCUTMENU <12>:** 0
> **Command:**

And the menu will no longer appear when a tool is in action.

Fig. 4.16 The *right-click* menu

Setting Layers (see also page 100)

1. *Left-click* on the **Layer Properties Manager** tool icon in the **Layers** toolbar (Fig. 4.17). The **Layer Properties Manager** dialog appears on screen (Fig. 4.18).
2. *Click* the **New Layer** icon. A new layer appears. Over the name **Layer1** *enter* **Centre**.
3. Repeat step **2** four times and make four more layers titled **Construction**, **Dimensions**, **Hidden** and **Text**.

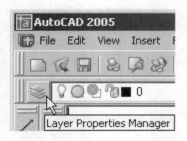

Fig. 4.17 The **Layer Properties Manager** tool icon in the **Layers** toolbar

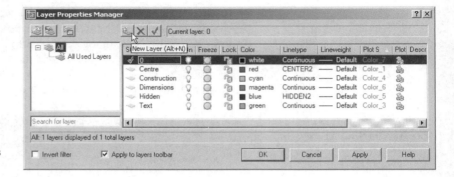

Fig. 4.18　The **Layer Properties Manager** dialog

4. *Click* against one of the squares under the **Color** column of the dialog. The **Select Color** dialog appears (Fig. 4.19). *Double-click* on one of the colours in the **Index Color** squares. The selected colour appears against the layer name in which the square was selected. Repeat until all five new layers have a colour.

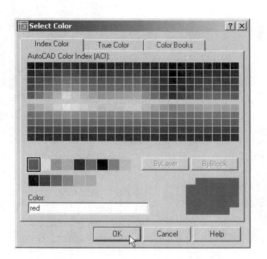

Fig. 4.19　The **Select Color** dialog

5. *Click* on the name **Continuous** against the layer name **Centre**. The **Select Linetype** dialog appears. *Click* its **Load** . . . button and from the **Load or Reload Linetypes** dialog *double-click* **CENTER2**. The dialog disappears and the name appears in the **Select Linetype** dialog. *Click* the **OK** button and the linetype **CENTER2** appears against the layer **Centre**.
6. Repeat against layer **Hidden** to load the linetype **HIDDEN2**.

Saving the template file

1. *Left-click* **Save As** . . . in the **File** drop-down menu.
2. In the **Save Drawing As** dialog which comes on screen (Fig. 4.20), *click* the arrow to the right of the **Files of type** field and in the popup list associated with the field *click* on **AutoCAD Drawing Template (*.dwt)**. The list of template files in the **AutoCAD 2005/Template** directory appears in the file list.

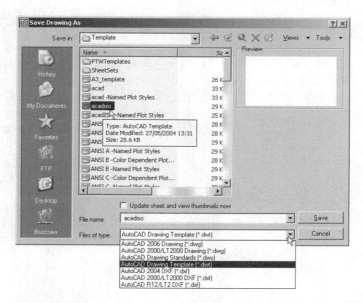

Fig. 4.20 Saving the template to the name **acadiso.dwt**

3. *Click* on **acadiso** in the file list, followed by a *click* on the **Save** button.
4. A **Template Description** dialog appears. Make *entries* as indicated in Fig. 4.21, making sure that **Metric** is chosen from the popup list.

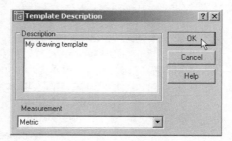

Fig. 4.21 The **Template Description** dialog

The template can now be saved and can be opened for the construction of drawings as needed. When AutoCAD 2005 is opened the template **acadiso.dwt** fills the drawing area.

Note

Please remember that if others are using the computer, it is advisable to save the template to a name of your own choice.

Another template

A template A4_template.dwt (Fig. 4.24)

In the **Select Template** dialog a *click* on any of the file names, causes a preview of the template to appear in the **Preview** box of the dialog, unless the template is free of information as in **acadiso.dwt**. To construct another template based on the **acadiso.dwt** template which includes a title block and other information:

Fig. 4.22 The **Set Current** icon in the **Layer Properties Manager**

1. From the **Select Template** dialog select the template just completed.
2. Make a new layer **Vports** of colour **Green**. Make this layer current with a *click* on the **Set Current** icon (Fig. 4.22).
3. *Click* the **Layout1** tab below the AutoCAD drawing area. The screen is now in a **Paper Space** setting.
4. *Click* the arrow at the right of the **Layers** field and in the popup list showing the layers which appears, *click* the **Turn a layer On or Off** icon against the layer name **Vports** (Fig. 4.23).

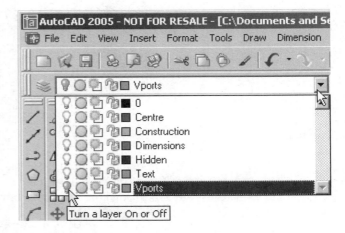

Fig. 4.23 Turn layer **Vports** off

5. Construct the border lines as shown in Fig. 4.24. To add the text *enter* **dt** (**Dynamic Text**) at the command line and follow the prompts which appear in the command window.
6. It is suggested this template be saved as a **Paper Space** template with the name **A3_template.dwt**.

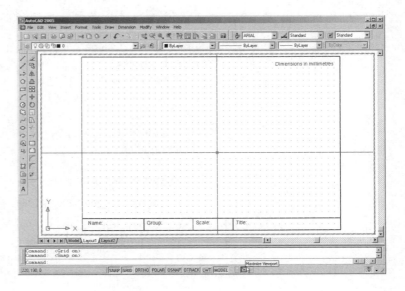

Fig. 4.24 The **A4_template.dwt**

Notes

1. The outline for this template is a pline from **0,290** to **420,290** to **420,0** to **0,0** to **290,0** and of width **0.5**.
2. The upper line of the title block is a pline from **0,20** to **420,20**.
3. The template was saved as a **Paper Space** file and maximised to fill the drawing area with a *click* on the **Maximize** button in the status bar, followed by a **Zoom Extents**.
4. Drawings cannot be constructed in this template (it being a **Pspace** file), unless the **MODEL** button in the status bar is **ON**.
5. **Pspace** is two-dimensional.
6. Further uses for **Layouts** and **Pspace** are given in Chapter 20.

Revision notes

1. The **Zoom** tools are important in that they allow even the smallest parts of drawings to be examined and, if necessary, amended or modified.
2. The zoom tools can be called from icons in the **Standard** or **Zoom** toolbars, or by *entering* **z** or **zoom** at the command line. The easiest is to *enter* **z** at the command line followed by a *right-click*.
3. There are four methods of calling tools for use – selecting a tool icon from a toolbar; *entering* the name of a tool in full at the command line; *entering* an abbreviation for a tool; selecting from a drop-down menu.
4. When constructing large drawings, the **Pan** tool and the **Aerial View** window are of value for allowing work to be carried out in any part of a drawing, while showing the whole drawing in the **Aerial View** window.
5. An A3 sheet of paper is 420 mm by 297 mm. If a drawing constructed in the template **acadiso.dwt** is printed/plotted full size (scale 1:1), each unit in the drawing will be 1 mm in the print/plot.
6. When limits are set it is essential to call **Zoom** followed by **a** (All) to ensure that the limits of the drawing area are as set.
7. If the *right-click* menu appears when using tools, the menu can be aborted by setting the **SHORTCUTMENU** variable to **0**.

CHAPTER 5

The Modify tools

Aim of this chapter

To describe the uses of tools from the **Modify** toolbar.

Introduction

The **Modify** tools are among the most frequently used tools of AutoCAD 2005. *Drag* the **Modify** toolbar (Fig. 5.1) from its usual position into the drawing area.

Fig. 5.1 The **Modify** toolbar

The **Erase** tool was described in Chapter 2. Examples of the other tools other than the **Explode** follow. See Chapter 9 for **Explode**.

The Copy Object tool

First example – Copy (Fig. 5.4)

1. Construct Fig. 5.3 using **Polyline**. Do not include the dimensions.
2. Call the **Copy Object** tool – either *left-click* on its tool icon in the **Modify** toolbar (Fig. 5.2), or *pick* **Copy** from the **Modify** drop-down

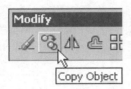

Fig. 5.2 The **Copy Object** tool icon from the **Modify** toolbar

Fig. 5.3 First example – **Copy Object** – outlines

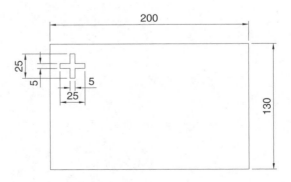

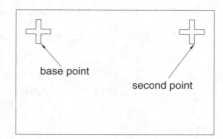

Fig. 5.4 First example – **Copy Object**

menu, or *enter* **cp** or **copy** at the command line. The command line shows:

Command:_copy
Select objects: *pick* the cross **1 found**
Select objects: *right-click*
Specify base point or displacement: end **of** *pick*
Specify second point of displacement or <use first point as displacement>: *pick*
Specify second point of displacement: *right-click*
Command:

First example – Multiple copy (Fig. 5.5)

1. Erase the copied object.
2. Call the **Copy Object** tool. The command line shows:

Command:_copy
Select objects: *pick* **1 found**
Select objects: *right-click*
Specify base point or displacement:
Specify base point: *pick*
Specify second point of displacement: *pick*
Specify second point of displacement or <use first point as displacement>: *pick*
Specify second point of displacement: *pick*
Specify second point of displacement: *right-click*
Command:

The result is shown in Fig. 5.5.

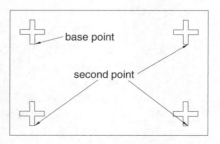

Fig. 5.5 First example – **Copy Object** – Multiple

Fig. 5.6 The **Mirror** tool icon
from the **Modify** toolbar

The Mirror tool

First example – Mirror (Fig. 5.8)

1. Construct the outline in Fig. 5.7 using **Line** and **Arc**.
2. Call the **Mirror** tool – either *left-click* on its tool icon in the **Modify** toolbar (Fig. 5.6), or *pick* **Mirror** from the **Modify** drop-down menu, or *enter* **mi** or **mirror** at the command line. The command line shows:

Command:_mirror
Select objects: *pick* first corner **Specify opposite corner:** *pick* **7 found**
Select objects: *right-click*
Specify first point of mirror line: end **of** *pick*
Specify second point of mirror line: end **of** *pick*
Delete source objects [Yes/No] <N>: *right-click*
Command:

The result is shown in Fig. 5.8.

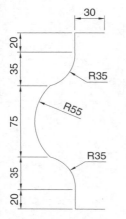

Fig. 5.7 First example –
Mirror – outline

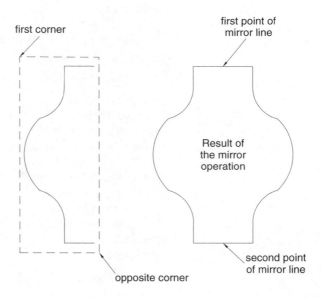

Fig. 5.8 First example – **Mirror**

Second example – Mirror (Fig. 5.9)

1. Construct the outline shown in the dimensioned polyline in Fig. 5.9.
2. Call **Mirror** and using the tool three times complete the given outline. The two points shown in Fig. 5.9 are to mirror the right-hand side of the outline.

Fig. 5.9 Second example –
Mirror

Third example – Mirror (Fig. 5.10)

If text is involved when using the **Mirror** tool, the set variable **MIRRTEXT** must be set correctly. To set the variable:

Command: mirrtext
Enter new value for MIRRTEXT <1>: 0
Command:

Fig. 5.10 Third example – **Mirror**

If set to **0**, text will mirror without distortion. If set to **1**, text will read backwards as indicated in Fig. 5.10.

The Offset tool

Example – Offset (Fig. 5.13)

1. Construct the four outlines shown in Fig. 5.12.
2. Call the **Offset** tool – either *left-click* its tool icon in the **Modify** toolbar (Fig. 5.11), or *pick* the tool name in the **Modify** drop-down menu, or *enter* **o** or **offset** at the command line. The command line shows:

Fig. 5.11 The **Offset** tool icon from the **Modify** toolbar

Command:_offset
Specify offset distance or [Through]: 10
Select object to offset or <exit>: *pick* drawing **1**
Specify point on side to offset: *pick* inside the rectangle
Select object to offset or <exit>: *right-click*
Command:

3. Repeat for drawings **2**, **3** and **4** in Fig. 5.12 as shown in Fig. 5.13.

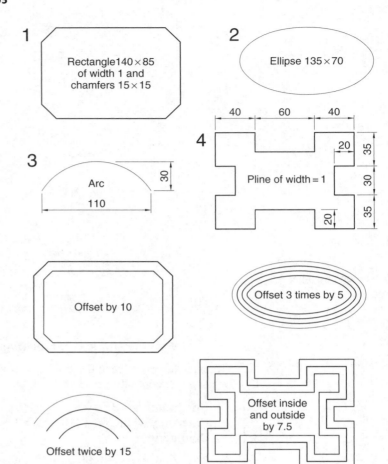

Fig. 5.12 First example – **Offset** – outlines

Fig. 5.13 Example – **Offset**

The Array tool

Arrays can be in either a **Rectangular** form or a **Polar** form as shown in the examples below.

First example – Rectangular Array (Fig. 5.17)

1. Construct the drawing in Fig. 5.14.
2. Call the **Array** tool – either *left-click* the **Array** tool icon in the **Modify** toolbar (Fig. 5.15), or *pick* **Array** . . . from the **Modify** drop-down menu, or *enter* **ar** or **array** at the command line. No matter which method is used the **Array** dialog appears (Fig. 5.16).
3. Make settings in the dialog:
   Rectangular Array radio button set on (dot in button).
   Row field – *enter* **5**
   Column field – *enter* **6**
   Row offset field – *enter* **−50** (note the minus sign)
   Column offset field – *enter* **50**

Fig. 5.14 First example – **Array** – drawing to be arrayed

Fig. 5.15 The **Array** tool icon from the **Modify** toolbar

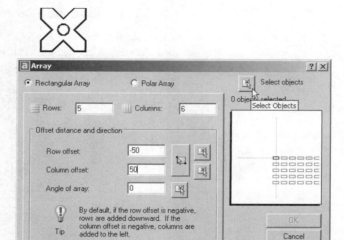

Fig. 5.16 First example – the
Array dialog

4. *Click* the **Select objects** button and the dialog disappears. Window the drawing. The dialog reappears.
5. *Click* the **Preview<** button. The dialog disappears and the array appears on screen with a warning dialog in the centre of the array (Fig. 5.17).
6. If satisfied *click* the **Accept** button. If not *click* the **Modify** button and make revisions to the **Array** dialog fields.

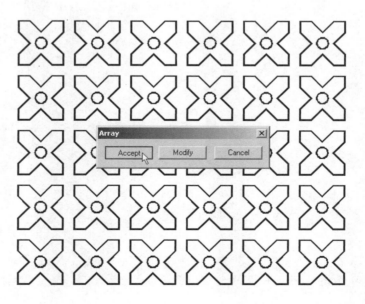

Fig. 5.17 First example – **Array**

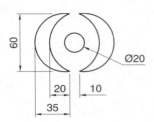

Fig. 5.18 Second example –
Array – drawing to be arrayed

Second example – Polar Array (Fig. 5.21)

1. Construct the drawing in Fig. 5.18.
2. Call **Array**. The **Array** dialog appears. Make settings as shown in Fig. 5.19.

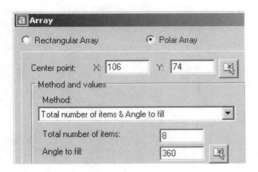

Fig. 5.19 Second example – the
setting in the **Array** dialog

3. *Click* the **Select objects** button of the dialog and window the drawing. The dialog returns to screen. *Click* the **Pick Center Point** button (Fig. 5.20) and when the dialog disappears, *pick* a centre point for the array.

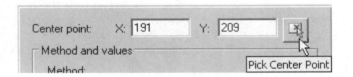

Fig. 5.20 Second example –
Array – the **Pick Center Point**
button

4. The dialog reappears. *Click* its **Preview<** button and when the array appears with its warning dialog, if satisfied with the result, *click* the **Accept** button of this dialog.

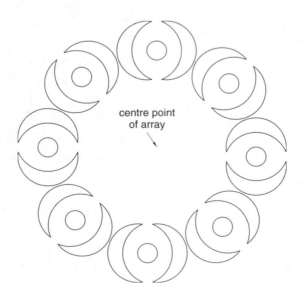

centre point
of array

Fig. 5.21 Second example – **Polar
Array**

The Move tool

Example – Move (Fig. 5.24)

1. Construct the drawing in Fig. 5.22.

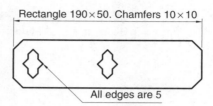
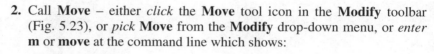

Rectangle 190×50. Chamfers 10×10

All edges are 5

Fig. 5.22 Example –
Move – drawing

Move

Fig. 5.23 The **Move** tool from
the **Modify** toolbar

2. Call **Move** – either *click* the **Move** tool icon in the **Modify** toolbar
(Fig. 5.23), or *pick* **Move** from the **Modify** drop-down menu, or *enter*
m or **move** at the command line which shows:

Command:_move
Select objects: *pick* the shape in the middle of the drawing
Select objects: *right-click*
Specify base point or displacement: *pick*
Specify second point of displacement: *pick*
Command:

The result is given in Fig. 5.24.

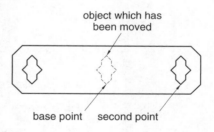

object which has
been moved

base point   second point

Fig. 5.24 Example – **Move**

The Rotate tool

When using the **Rotate** tool remember the default rotation of objects
within AutoCAD 2005 is counterclockwise (anticlockwise).

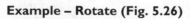

Rotate

Fig. 5.25 The **Rotate** tool icon
from the **Modify** toolbar

Example – Rotate (Fig. 5.26)

1. Construct drawing **1** of Fig. 5.26 with **Polyline**. Copy the drawing **1**
three times – shown as drawings **2**, **3** and **4** in Fig. 5.26.
2. Call **Rotate** – either *left-click* its tool icon in the **Modify** toolbar
(Fig. 5.25), or *pick* **Rotate** from the **Modify** drop-down menu, or *enter* **ro**
or **rotate** at the command line. The command line shows:

Command:_rotate
Current positive angle in UCS: ANGDIR = counterclockwise
ANGBASE = 0

Select objects: window the drawing **3 found**
Select objects: *right-click*
Specify base point: *pick*
Specify rotation angle: 45
Command:

And the first copy rotates through the specified angle.

3. Repeat for drawings **3** and **4** rotating through angles as shown in Fig. 5.26.

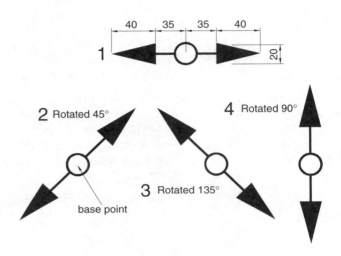

Fig. 5.26 Example – **Rotate**

The Scale tool

Examples – Scale (Fig. 5.28)

1. Using the **Rectangle** and **Polyline** tools, construct drawing **1** of Fig. 5.28. The **Rectangle** fillets are R10. The line width of all parts is **1**. Copy the drawing three times to give drawings **2**, **3** and **4**.
2. Call **Scale** – either *left-click* its tool icon in the **Modify** toolbar (Fig. 5.27), or *pick* **Scale** from the **Modify** drop-down menu, or *enter* **sc** or **scale** at the command line which then shows:

Fig. 5.27 The **Scale** tool icon from the **Modify** toolbar

Command:_scale
Select objects: window drawing **3**
Select objects: *right-click*
Specify base point: *pick*
Specify scale factor or [Reference]: 0.75
Command:

3. Repeat for the other two drawings **2** and **4** scaling to the scales given with the drawings.

The results are shown in Fig. 5.28.

Fig. 5.28 Examples – **Rotate**

The Trim tool

This tool is one which will be frequently used for the construction of drawings.

First example – Trim (Fig. 5.30)

1. Construct the drawing labelled **Original drawing** in Fig. 5.30 with **Circle** and **Line**.
2. Call **Trim** – either *left-click* its tool icon in the **Modify** toolbar (Fig. 5.29), or *pick* **Trim** from the **Modify** drop-down menu, or *enter* **tr** or **trim** at the command line, which then shows:

Fig. 5.29 The **Trim** tool icon from the **Modify** toolbar

Command:_trim
Current settings: Projection UCS. Edge=None
Select cutting edges: *pick* the left-hand circle **1 found**
Select objects: *right-click*
Select objects to trim or shift-select to extend or [Project/Edge/ Undo]: *pick* one of the objects
Select objects to trim or shift-select to extend or [Project/Edge/ Undo]: *pick* the second of the objects
Select objects to trim or shift-select to extend or [Project/Edge/ Undo]: *right-click*
Command:

3. This completes the **First stage** as shown in Fig. 5.30. Repeat the **Trim** sequence for the **Second stage**.
4. The **Third stage** drawing of Fig. 5.30 shows the result of the trims at the left-hand end of the drawing.
5. Repeat for the right-hand end. The final result is shown in the drawing labelled **Result** in Fig. 5.30.

Second example – Trim (Fig. 5.31)

1. Construct the left-hand drawing of Fig. 5.31.
2. Call **Trim**. The command line shows:

Command:_trim
Current settings: Projection UCS. Edge=None

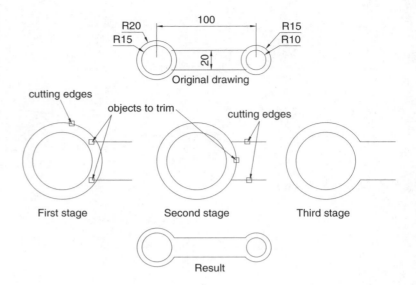

Fig. 5.30 First example – **Trim**

> **Select cutting edges:** *pick* the left-hand arc **1 found**
> **Select objects:** *right-click*
> **Select objects to trim or shift-select to extend or [Project/Edge/ Undo]:** e (Edge)
> **Enter an implied edge extension mode [Extend/No extend] <No extend>:** e (Extend)
> **Select objects to trim:** *pick*
> **Select objects to trim:** *pick*
> **Select objects to trim:** *right-click*
> **Command:**

3. Repeat for the other required trims. The result is given in Fig. 5.31.

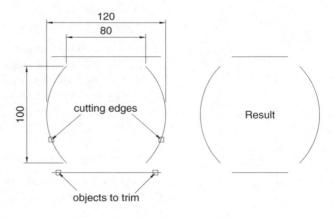

Fig. 5.31 Second example – **Trim**

The Stretch tool

Examples – Stretch (Fig. 5.33)

As its name implies the **Stretch** tool is for stretching drawings or parts of drawings. The action of the tool prevents it from altering the shape of circles in any way. Only **crossing** or **polygonal** windows can be used to determine the part of a drawing which is to be stretched.

1. Construct the drawing labelled **Original** in Fig. 5.33, but do not include the dimensions. Use the **Circle**, **Arc**, **Trim** and **Polyline Edit** tools. The resulting outlines are plines of width = 1. With the **Copy Object** tool make two copies of the drawing.

   Note

   In each of the three examples in Fig. 5.33, the broken lines represent the crossing windows required when **Stretch** is used.

Stretch

Fig. 5.32 The **Stretch** tool icon from the **Modify** toolbar

2. Call the **Stretch** tool – either *click* on its tool icon in the **Modify** toolbar (Fig. 5.32), or *pick* its name in the **Modify** drop-down menu, or *enter* **s** or **stretch** at the command line which shows:

   Command:_stretch
   Select objects to stretch by crossing-window or crossing-polygon.
   Select objects: *enter* **c** *right-click*
   Specify first corner: *pick* **Specify opposite corner:** *pick* **1 found**
   Select objects: *right-click*
   Specify base point or displacement: *pick* beginning of arrow
   Specify second point of displacement or <use first point as displacement>: *drag* in the direction of the arrow to the required second point and *right-click*
   Command:

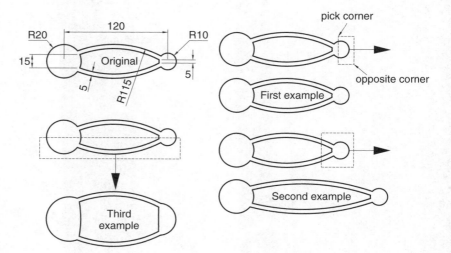

Fig. 5.33 Examples – **Stretch**

Notes

1. When circles are windowed with the crossing-window no stretching can take place. This is why, in the case of the first example in Fig. 5.33, when the **second point of displacement** was *picked*, there was no result – the outline did not stretch.
2. Care must be taken when using this tool as unwanted stretching can occur.

The Break tool

Examples – Break (Fig. 5.35)

1. Construct the rectangle, arc and circle (Fig. 5.35).
2. Call **Break** – either *click* its tool icon in the **Modify** toolbar (Fig. 5.34), or *click* **Break** from the **Modify** drop-down menu, or *enter* **br** or **break** at the command line which shows:

For drawings 1 and 2

Command:_break Select object: *pick* at the point
Specify second break point or [First point]: *pick*
Command:

For drawing 3

Command:_break Select object: *pick* at the point
Specify second break point or [First point]: *enter* **f** *right-click*
Specify first break point: *pick*
Specify second break point: *pick*
Command:

The results are shown in Fig. 5.35.

Fig. 5.34 The **Break** tool icon from the **Modify** toolbar

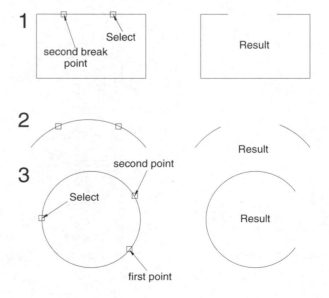

Fig. 5.35 Examples – **Break**

Note

Remember the default rotation of AutoCAD 2005 is counterclockwise. This applies to the use of the **Break** tool.

The Extend tool

Examples – Extend (Fig. 5.37)

1. Construct plines and a circle as shown in the left-hand drawings of Fig. 5.37.
2. Call **Extend** – either *click* the **Extend** tool in the **Modify** toolbar (Fig. 5.36), or *pick* **Extend** from the **Modify** drop-down menu, or *enter* **ex** or **extend** at the command line which then shows:

Command:_extend
Current settings: Projection=UCS Edge=Extend
Select boundary edges: *pick*
Select objects: *right-click*
Select object to extend or shift-select to trim or [Project/Edge/ Undo]: *pick*
Repeat for each object to be extended. Then:
Select object to extend or shift-select to trim or [Project/Edge/ Undo]: *right-click*
Command:

Fig. 5.36 The **Extend** tool icon from the **Modify** toolbar

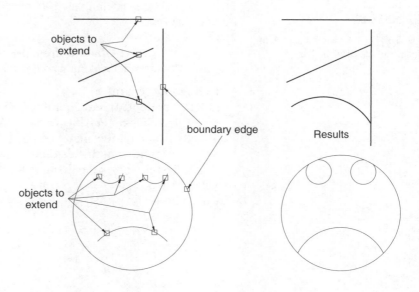

Fig. 5.37 Examples – **Extend**

Note

Observe the similarity of the **Extend** and **No extend** prompts with those of the **Trim** tool.

The Chamfer and Fillet tools

There are similarities in the prompt sequences for these two tools. The major differences are that two settings (**Dist1** and **Dist2**) are required for a chamfer, but only one (**Radius**) for the fillet. The basic prompts for both are:

Chamfer

Command:_chamfer
(TRIM mode) Current chamfer Dist1=1, Dist2=1
Select first line or [Polyline/Distance/Angle/Trim/Method/mUltiple]:
    enter **d** (Distance) *right-click*
Specify first chamfer distance <1>: 10
Specify second chamfer distance <10>: *right-click*
Command:

Fillet

Command:_fillet
Current settings: Mode=TRIM, Radius=1
Select first object or [Polyline/Radius/Trim/mUltiple]: *enter* **r** (Radius)
    right-click
Specify fillet radius <1>: 15
Command:

Fig. 5.38 The **Chamfer** tool icon from the **Modify** toolbar

Examples – Chamfer (Fig. 5.39)

1. Construct three rectangles 100 by 60 using either the **Line** or the **Polyline** tool.
2. Call **Chamfer** – either *click* its tool icon in the **Modify** toolbar (Fig. 5.38), or *pick* **Chamfer** from the **Modify** drop-down menu, or *enter* **cha** or **chamfer** at the command line which then shows:

Command:_chamfer
(TRIM mode) Current chamfer Dist1=1, Dist2=1
Select first line or [Polyline/Distance/Angle/Trim/Method/mUltiple]: d
Specify first chamfer distance <1>: 10
Specify second chamfer distance <10>: *right-click*
Select first line: *pick* the first line for the chamfer
Select second line: *pick*
Command:

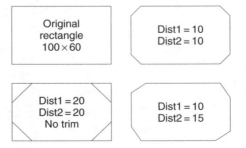

Fig. 5.39 Examples – **Chamfer**

The other two rectangles are chamfered in a similar manner except that the **No trim** prompt is brought into operation with the bottom left-hand example.

Examples – Fillet (Fig. 5.41)

1. Construct three rectangles as for the **Chamfer** examples.
2. **Call** Fillet – either *click* its tool icon in the **Modify** toolbar (Fig. 5.40), or *pick* **Fillet** from the **Modify** drop-down menu, or *enter* **f** or **fillet** at the command line which then shows:

   Command:_fillet
   Current settings: Mode=TRIM, Radius=1
   Select first object or [Polyline/Radius/Trim/mUltiple]: r (Radius)
   Specify fillet radius <1>: 15
   Select first object: *pick*
   Select second object: *pick*
   Command:

Three examples are given in Fig. 5.41.

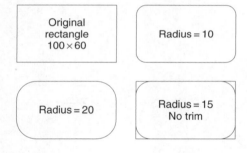

Fig. 5.40 The **Fillet** tool icon from the **Modify** toolbar

Fig. 5.41 Examples – **Fillet**

Revision notes

1. The **Modify** tools are among the most frequently used tools in Auto-CAD 2005.
2. The abbreviations for the **Modify** tools are:
   Copy Object – cp
   Mirror – mi
   Offset – o
   Array – ar
   Move – m
   Rotate – ro
   Scale – sc
   Stretch – s
   Trim – tr
   Extend – ex
   Break – br
   Chamfer – cha
   Fillet – f

3. There are two other tools in the **Modify** toolbar: **Erase** – some examples were given in Chapter 2; and **Explode** – further details of this tool will be given in Chapter 9.
4. When using **Mirror**, if text is part of the area to be mirrored, the set variable **MIRRTEXT** will require setting – to either **1** or **0**.
5. With **Offset** the **Through** prompt can be answered by *clicking* two points in the drawing area – the distance of the desired offset distance.
6. **Polar Arrays** can be arrays around any angle set in the **Angle of array** field of the **Array** dialog.
7. When using **Scale**, it is advisable to practise the **Reference** prompt.
8. The **Trim** tool in either its **Trim** or its **No trim** modes is among the most useful tools in AutoCAD 2005.
9. When using **Stretch**, circles are unaffected by the stretching.

Exercises

1. Construct the Fig. 5.42. All parts are plines of width = 0.7 with corners filleted R10. The long strips have been constructed using **Circle, Polyline, Trim** and **Polyline Edit**. Construct one strip and then copy it using **Copy Object**.

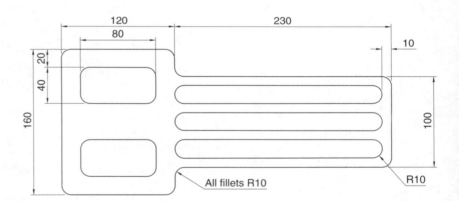

Fig. 5.42 Exercise 1

2. Construct the drawing in Fig. 5.43. All parts of the drawing are plines of width = 0.7. The circles are arrayed from a single circle using the **Array** tool.
3. Using the tools **Polyline, Circle, Trim, Polyline Edit, Mirror** and **Fillet**, construct the drawing in Fig. 5.44.
4. Construct the circles and lines shown in Fig. 5.45. Using **Offset** and the **Ttr** prompt of the **Circle** tool, followed by **Trim**, construct one of the outlines arrayed within the outer circle. Then with **Polyline Edit** change the lines and arcs into a pline of width = 0.3. Finally array the outline twelve times around the centre of the circles to produce Fig. 5.46.

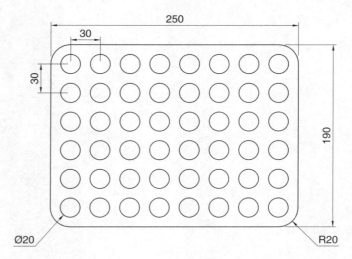

Fig. 5.43 Exercise 2

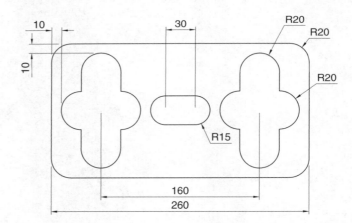

Fig. 5.44 Exercise 3

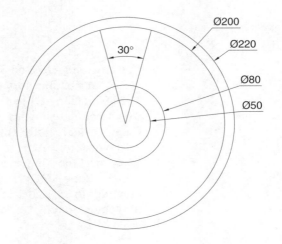

Fig. 5.45 Exercise 4 – circles
and lines on which the exercise
is based

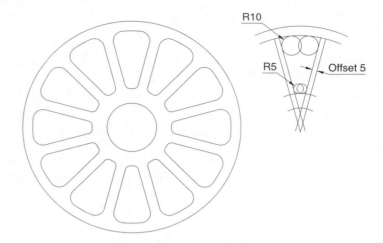

Fig. 5.46 Exercise 4

5. Construct the arrow (Fig. 5.47). Array the arrow around the centre of its circle eight times to produce the right-hand drawing of Fig. 5.47.

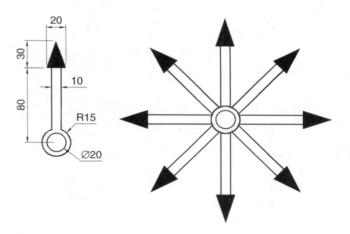

Fig. 5.47 Exercise 5

6. Construct the left-hand drawing of Fig. 5.48. Then with **Move**, move the central outline to the top left-hand corner of the outer outline. Then with **Copy Object** make copies to the other corners.

7. Construct the drawing in Fig. 5.49 and make two copies using **Copy Object**. With **Rotate** rotate each of the copies to the angles as shown.

8. Construct the dimensioned drawing of Fig. 5.50. With **Copy Object** copy the drawing. Then with **Scale** scale the original drawing to a scale of 0.5, followed by **Rotate** to rotate the drawing through the angle as shown. Finally scale the copy to a scale of 2:1.

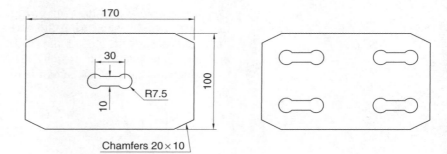

Fig. 5.48 Exercise 6

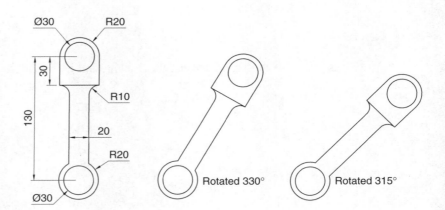

Fig. 5.49 Exercise 7

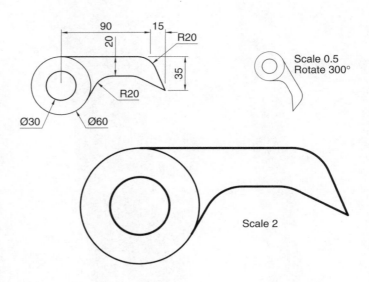

Fig. 5.50 Exercise 8

9. Construct the left-hand drawing of Fig. 5.51. Include the dimensions in your drawing. Then, using the **Stretch** tool, stretch the drawing, including its dimensions to the sizes as shown in the right-hand drawing. Note that the dimensions change as the drawing is stretched. The dimensions are said to be **associative** (See Chapter 6).

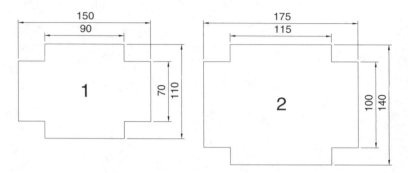

Fig. 5.51 Exercise 9

10. Construct the drawing in Fig. 5.52. All parts of the drawing are plines of width = 0.7. The setting in the **Array** dialog is to be **180** in the **Angle of array** field.

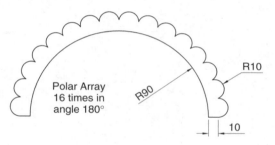

Fig. 5.52 Exercise 10

CHAPTER 6

Dimensions and Text

Aims of this chapter

1. To describe the variety of methods of dimensioning drawings.
2. To describe methods of adding text to drawings.

Introduction

We have already set a dimension style (**my_style**) in the **acadiso.dwt** template, so we can now commence adding dimensions to drawings using this dimension style. This style name appears in the **Styles** toolbar because AutoCAD 2005 opens with the **acadiso.dwt** template.

The Dimension tools

Right-click in any toolbar on screen and select **Dimension** from the menu which appears. The toolbar appears (Fig. 6.1). Two tooltips are shown – **Linear Dimension** and **Dimension Style**. *Click* on the icon of the latter and the **Dimension Style Manager** appears. Note the name of the dimension style (**my_style**) that appears in the toolbar. If several dimension styles have been saved, a popup list from the field showing all style names will appear allowing a choice of style to be made.

Fig. 6.1 The **Dimension** toolbar

Adding dimensions using the tools

First example – Linear Dimension (Fig. 6.2)

1. Construct a rectangle 180×110 using the **Polyline** tool.
2. *Left-click* on the **Linear Dimension** tool icon in the **Dimension** toolbar (Fig. 6.1). The command line shows:

Command:_dimlinear
Specify first extension line origin or [select object]: *pick*

Specify second extension line origin: *pick*
**Specify dimension line location or [Mtext/Text/Angle/Horizontal/
    Vertical/Rotated]:** *pick*
Dimension text = 180
Command:

Fig. 6.2 shows the 180 dimension. Follow exactly the same procedure for the 110 dimension.

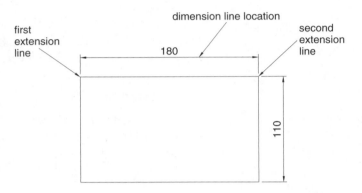

Fig. 6.2 First example – **Linear Dimension**

Notes

1. If necessary use **Osnaps** to locate the extension line locations.
2. The prompt **Specify first extension line origin or [select object]:** allows the line being dimensioned to be selected.

Second example – Aligned Dimension (Fig. 6.4)

1. Construct the outline in Fig. 6.4 using the **Line** tool.
2. *Left-click* the **Aligned Dimension** tool icon (Fig. 6.3) and dimension the outline. The prompts and replies are similar to the first example.

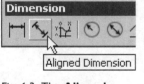

Fig. 6.3 The **Aligned Dimension** tool icon in the **Dimension** toolbar

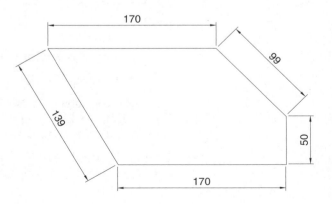

Fig. 6.4 Second example – **Aligned Dimension**

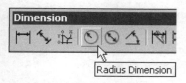

Fig. 6.5 The **Radius Dimension** tool from the **Dimension** toolbar

Third example – Radius Dimension (Fig. 6.6)

1. Construct the outline in Fig. 6.6 using the **Line** and **Fillet** tools.
2. *Left-click* the **Radius Dimension** tool icon in the **Dimension** toolbar (Fig. 6.5). The command line shows:

Command:_dimradius
Select arc or circle: *pick* one of the arcs
Dimension text = 30
Specify dimension line location or [Mtext/Text/Angle]: *pick*
Command:

3. Continue dimensioning the outline as shown in Fig. 6.6.

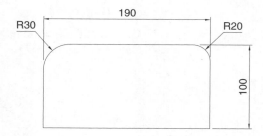

Fig. 6.6 Third example – **Radius Dimension**

Notes

1. At the prompt [**Mtext/Text/Angle**]:
   (a) If a **t** (Text) is *entered*, another number can be *entered*, but remember if the dimension is a radius the letter **R** must be *entered* as a prefix to the new number.
   (b) If the response is **a** (Angle), and an angle number is *entered*, the text for the dimension will appear at an angle. Fig. 6.7 shows a radius dimension *entered* at an angle of 45°.
   (c) If the response is **m** (Mtext) the **Text Formatting** dialog appears together with a box in which new text can be *entered*. See page 93.
2. Dimensions added to a drawing using other tools from the toolbar should be practised.

Fig. 6.7 A radius dimension at an angle of 45°

Adding dimensions from the command line

Fig. 6.8 shows all the tooltips for the tools in the **Dimension** toolbar. Not all of these have been shown in the examples above. Some operators may

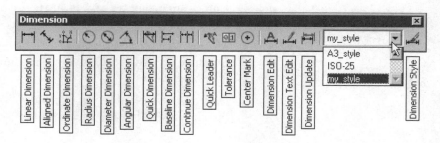

Fig. 6.8 All tools in the **Dimension** toolbar

prefer *entering* dimensions from the command line. This involves *entering* abbreviations for the required dimension such as:

> For **Linear Dimension** – **hor** (horizontal) or **ve** (vertical);
> For **Aligned Dimension** – **al**;
> For **Radius Dimension** – **ra**;
> For **Diameter Dimension** – **d**;
> For **Angular Dimension** – **an**;
> For **Dimension Text Edit** – **te**;
> For **Quick Leader** – **l**;
> And to exit from the dimension commands – **e** (Exit).

> ### First example – hor and ve (Horizontal and Vertical) – (Fig. 6.10)

1. Construct the outline in Fig. 6.9 using the **Line** tool. Its dimensions are shown in Fig. 6.10.

Fig. 6.9 First example – outline for dimensions

2. At the command line *enter* **dim**. The command line will show:

> **Command:** *enter* **dim** *right-click*
> **Dim:** *enter* **hor** (horizontal) *right-click*
> **Specify first extension line origin or <select object>:** *pick*
> **Specify second extension line origin:** *pick*
> **Specify dimension line location or [Mtext/Text/Angle]:** *pick*
> **Enter dimension text <50>:** *right-click*
> **Dim:** *right-click*
> **HORIZONTAL**
> **Specify first extension line origin or <select object>:** *pick*
> **Specify second extension line origin:** *pick*
> **Specify dimension line location or [Mtext/Text/Angle/Horizontal/**
>   **Vertical/Rotated]:** *pick*
> **Enter dimension text <140>:** *right-click*
> **Dim:** *right-click*

And the 50 and 140 horizontal dimensions are added to the outline.

3. Continue to add the right-hand 50 dimension. Then when the command line shows:

> **Dim:** *enter* **ve** (vertical) *right-click*
> **Specify first extension line origin or <select object>:** *pick*

Specify second extension line origin: *pick*
Specify dimension line location or [Mtext/Text/Angle/Horizontal/ Vertical/Rotated]: *pick*
Dimension text <20>: *right-click*
Dim: *right-click*
VERTICAL
Specify first extension line origin or <select object>: *pick*
Specify second extension line origin: *pick*
Specify dimension line location or [Mtext/Text/Angle/Horizontal/ Vertical/Rotated]: *pick*
Dimension text <100>: *right-click*
Dim: *enter* **e** (Exit) *right-click*
Command:

The result is shown in Fig. 6.10.

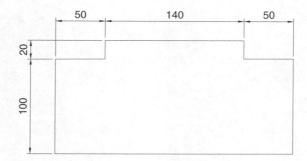

Fig. 6.10 First example – horizontal and vertical dimensions

Second example – an (Angular) – (Fig. 6.12)

1. Construct the outline in Fig. 6.11 – a pline of width = 1.

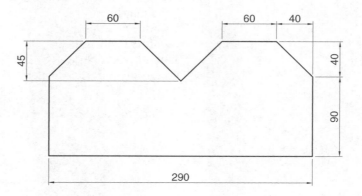

Fig. 6.11 Second example – outline for dimensions

2. At the command line:

Command: *enter* **dim** *right-click*
Dim: *enter* **an** *right-click*
Select arc, circle, line or <specify vertex>: *pick*

Select second line: *pick*
Specify dimension arc line location or [Mtext/Text/Angle]: *pick*
Enter dimension <90>: *right-click*
Enter text location (or press ENTER): *pick*
Dim:

And so on to add the other angular dimensions.
The result is given in Fig. 6.12.

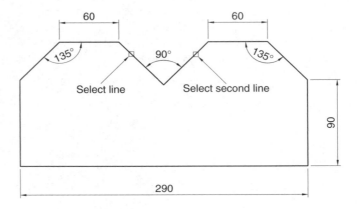

Fig. 6.12 Second example – **an** dimension

Third example – l (Leader) – (Fig. 6.14)

1. Construct Fig. 6.13.
2. At the command line:

Command: *enter* **dim** *right-click*
Dim: *enter* **l** (Leader) *right-click*
Leader start: *enter* **nea** (osnap nearest) *right-click* **to** *pick* one of the chamfer lines
To point: *pick*
To point: *pick*
To point: *right-click*
Dimension text <0>: *enter* **CHA 10×10** *right-click*
Dim: *right-click*

Continue to add the other leader dimensions (Fig. 6.14).

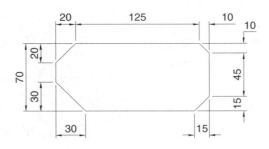

Fig. 6.13 Third example – outline for dimensioning

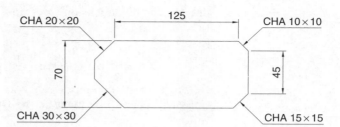

Fig. 6.14 Third example – **I**
dimensions

Fourth example – te (Dimension Text Edit) – (Fig. 6.16)

1. Construct Fig. 6.15.
2. At the command line:

Command: *enter* **dim** *right-click*
Dim: *enter* **te** (tedit) *right-click*
Select dimension: *pick* the dimension to be changed
Specify new location for text or [Left/Right/Center/Home/Angle]:
    either *pick* or *enter* a prompt's capital letter
Dim:

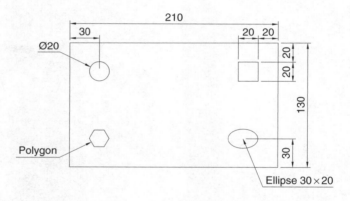

Fig. 6.15 Fourth example –
dimensioned drawing

The results as given in Fig. 6.16 show dimensions which have been
moved. The **210** dimension changed to the left-hand end of the dimension

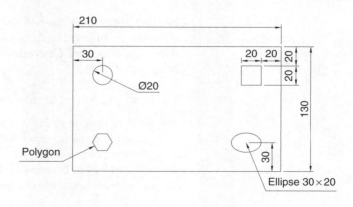

Fig. 6.16 Fourth example –
dimensions amended with **tedit**

line and the **30** dimension's position changed. Remainder of the two lines to be deleted.

Dimension tolerances

Before simple tolerances can be included with dimensions, new settings will need to be made in the **Dimension Style Manager** dialog as follows:

1. Open the dialog. The quickest way of doing this is to *enter* **d** at the command line followed by a *right-click*. This opens up the dialog.
2. *Click* the **Modify** . . . button of the dialog, followed by a *left-click* on the **Primary Units** tab and in the resulting sub-dialog make settings as shown in Fig. 6.17. Note the changes in the preview box of the dialog.

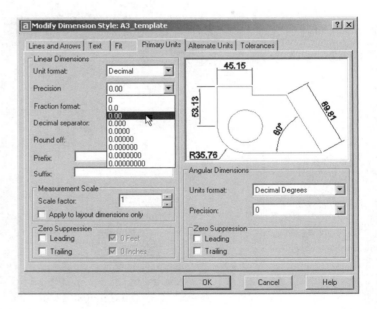

Fig. 6.17 The **Primary Units** sub-dialog of the **Dimension Style Manager**

3. *Click* the **Tolerances** tab and in the resulting sub-dialog, make settings as shown in Fig. 6.18. *Left-click* the **OK** button, then in the main dialog, *click* the **Set Current** button, followed by a *left-click* on the **Close** button.

Example – simple tolerances (Fig. 6.20)

1. Construct the outline in Fig. 6.19.
2. Dimension the drawing using either tools from the **Dimension** toolbar or by *entering* abbreviations at the command line. Because tolerances have been set in the **Dimension Style Manager** dialog, the toleranced dimensions will automatically be added to the drawing.

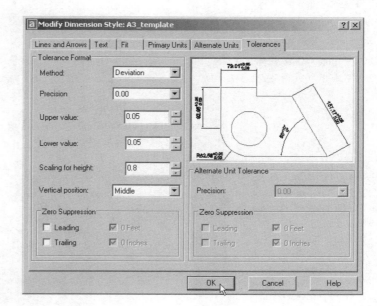

Fig. 6.18 The **Tolerances** sub-dialog of the **Dimension Style manager**

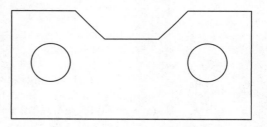

Fig. 6.19 Example – Simple tolerances – outline

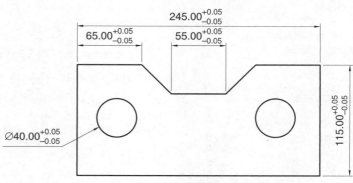

Fig. 6.20 Example – Simple tolerances – **Dynamic Text**

The dimensions in this drawing show tolerances

Example – Geometrical Tolerance (Fig. 6.26)

1. Construct the two rectangles with circles as in Fig. 6.21.
2. Add dimensions to the two circles.

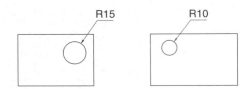

Fig. 6.21 Example – **Geometrical Tolerance** – dimensions to be toleranced

Fig. 6.22 The **Tolerance** tool icon in the **Dimension** toolbar

3. *Click* the **Tolerance** tool icon (Fig. 6.22). The **Geometrical Tolerance** dialog (Fig. 6.23) appears.

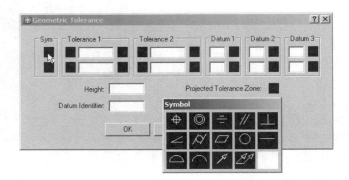

Fig. 6.23 The **Geometrical Tolerance** dialog and the **Symbol** sub-dialog

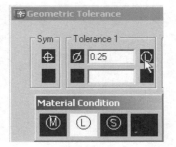

Fig. 6.24 The **Material Condition** dialog

4. In the dialog *click* the black box under **Sym**. The **Symbol** sub-dialog (Fig. 6.23) appears.
5. Still in the dialog *click* the left-hand black square under **Tolerance 1**. The **Material Condition** dialog appears (Fig. 6.24). *Click* **L**. The letter appears in the top right-hand square of the dialog.
6. *Enter* **0.05** in the **Tolerance 1** field (Fig. 6.25), followed by a *click* on the dialog's **OK** button. The geometrical tolerance appears. Move it to a position near the **R10** dimension in the drawing (Fig. 6.26).
7. Now add a geometrical tolerance to the **R15** dimension as shown in Fig. 6.26.

The meanings of the symbols

The **Material Condition** letters have the following meanings:

M – maximum amount of material.
L – least amount of material.
S – size within the limits.

Fig. 6.27 shows the meanings of the geometrical symbols.

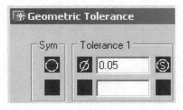

Fig. 6.25 *Enter* **0.05** in the **Tolerance 1** field

Fig. 6.26 Example – **Geometrical tolerances**

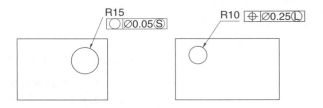

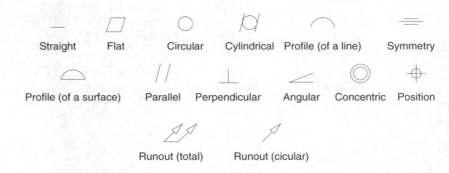

Fig. 6.27 The meanings of the symbols

Text

There are two main methods of adding text to drawings – **Dynamic Text** and **Multiline Text**.

Example – Dynamic Text (Fig. 6.20)

1. Open the drawing from the example on tolerances – Fig. 6.20.
2. At the command line *enter* **dt** (Dynamic Text) followed by a *right-click*:

Command: *enter* **dt** *right-click*
TEXT
Current text style "ARIAL" Text height: 8
Specify start point of text or [Justify/Style]: *pick*
Specify rotation angle of text <0>: *right-click*
Enter text: *enter* **The dimensions in this drawing show tolerances**
   press the **Return** key
Enter text: *press* the **Return** key
Command:

The result is given in Fig. 6.20.

Notes

1. It is essential that the **Return** key of the keyboard is pressed and **NOT** a *right-click* when *entering* text using the **Dynamic Text** tool.
2. At the prompt:

Specify start point of text or [Justify/Style]: *enter* **s** (Style) *right-click*
Enter style name or [?] <ARIAL>: *enter* **?** *right-click*
Enter text style(s) to list <*>: *right-click*

And an **AutoCAD Text Window** (Fig. 6.29) appears listing all the styles which have been selected in the **Text Style** (see page 53).
3. In order to select the required text style its name must be *entered* at the prompt:

Enter style name or [?] <ARIAL>: *enter* **romand** *right-click*

And the text *entered* will be in the **Romand** style of height **9**. But only if that style was selected in the **Text Style** dialog.

4. Fig. 6.28 shows some text from the styles from the **AutoCAD Text Window** (Fig. 6.29).

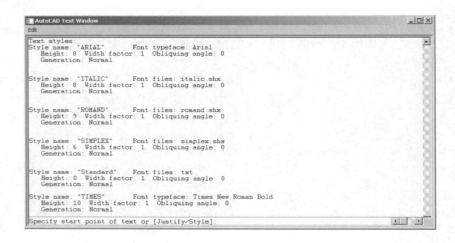

This is the TIMES text

This is ROMANC text

This is ROMAND text

This is STANDARD text

This is ITALIC text

This is ARIAL text

Fig. 6.28 Some text styles

Fig. 6.29 The **AutoCAD Text Window** already selected in the **Text Style** dialog

5. There are two types of text fonts available in AutoCAD 2005 – the **AutoCAD SHX** fonts and the **Windows True Type** fonts. The **ITALIC**, **ROMAND**, **ROMANS** and **STANDARD** styles shown in Fig. 6.28 are AutoCAD text fonts. The **TIMES** and **ARIAL** styles are **Windows True Type** fonts. Most of the **True Type** fonts can be *entered* in **Bold**, **Bold Italic**, **Italic** or **Regular** styles, but these variations are not possible with the AutoCAD fonts.
6. In the **Font name** popup list of the **Text Style** dialog, it will be seen that a large number of text styles are available to the AutoCAD 2005 operator. It is advisable to practise using a variety of these fonts to familiarise oneself with the text styles available with AutoCAD 2005.

Example – Multiline Text (Fig. 6.32)

1. Open the **A3_template.dwt** template.
2. Either *left-click* on the **Multiline Text** tool icon in the **Draw** toolbar (Fig. 6.30) or *enter* **t** at the command line:

Fig. 6.30 The **Multiline Text** tool icon from the **Draw** toolbar

Command:_mtext
Current text style: "ARIAL" Text height: 8

Specify first corner: *pick*

Specify opposite corner or [Height/Justify/Line spacing/Rotation/ Style/Width]: *pick*

And a box such as in Fig. 6.31 appears on screen. As soon as the **opposite corner** is *picked*, the **Text Formatting** dialog appears and the box changes as in Fig. 6.32. Text can now be *entered* as required within the box as indicated in this illustration.

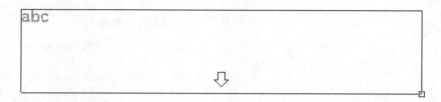

Fig. 6.31 The **Multiline Text** box

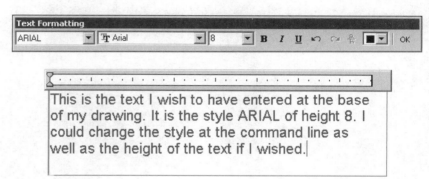

Fig. 6.32 Example – **Multiline Text** *entered* in the text box

When all the required text has been *entered left-click* the **OK** button at the top right-hand corner of the **Text Formatting** dialog.

3. Changes may be made to various aspects of the text being *entered*, making choices from the various popup lists in the **Text Formatting** dialog. These popups are shown in Fig. 6.33.

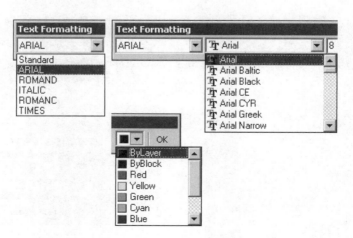

Fig. 6.33 The popups from the **Text Formatting** dialog

Symbols used in text

When text has to be added by *entering* letters and figures as part of a dimension, the following symbols must be used:

To obtain **Ø75** *enter %%c75*;
To obtain **55%** *enter 55%%%*;
To obtain **±0.05** *enter %%p0.05*;
To obtain **90°** *enter 90%%d*.

Checking spelling

There are two methods for the checking of spellings in AutoCAD 2005.

First example – the Spelling tool (Fig. 6.35)

1. *Enter* some badly spelt text as indicated in Fig. 6.35.
2. *Left-click* on **Edit** . . . from the **Text** sub-menu in the **Modify** drop-down menu (Fig. 6.34).

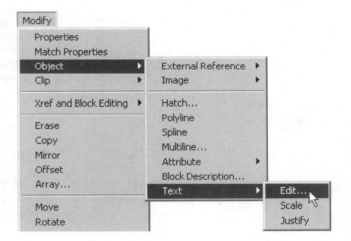

Fig. 6.34 Selecting the **Edit Text** dialog

3. *Left-click* on the text. The **Edit Text** dialog appears with the text to be edited in its **Text** field. Edit the text as if working in a word processing application and when satisfied *click* the **OK** button (Fig. 6.35).

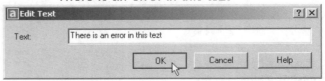

Fig. 6.35 First example – the **Edit Text** dialog

Second example – the Spelling tool (Fig. 6.36)

1. *Enter* some badly spelt text as indicated in Fig. 6.35.
2. Either *click* **Spelling** in the **Tools** drop-down menu (Fig. 6.37) or *enter* **spell** or **sp** at the command line.
3. *Click* the badly spelt text and *right-click*. The **Check Spelling** dialog appears (Fig. 6.36). Wrongly spelt words appear in the **Current word** field with words to replace them in the **Suggestions** field. Select the appropriate correct spelling as shown. Continue until all text is checked.

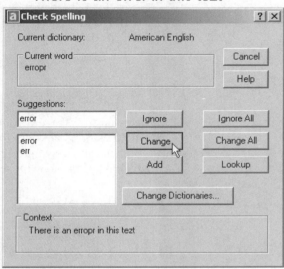

Fig. 6.36 Second example – the **Check Spelling** dialog

Fig. 6.37 **Spelling** in the **Tools** drop-down menu

Revision notes

1. In the **Line and Arrows** sub-dialog of the **Dimension Style Manager** dialog, **Lineweights** were set to **0.3** (page 54). If these lineweights are to show in the drawing area of AutoCAD 2005, the **LWT** button in the status bar must be set **ON**.
2. Dimensions can be added to drawings using the tools from the **Dimension** toolbar or by *entering* abbreviations for the tools at the command line.
3. It is usually advisable to use osnaps when locating points on a drawing for dimensioning.
4. The **Style** and **Angle** of the text associated with dimensions can be changed during the dimensioning process.
5. When wishing to add tolerances to dimensions it will probably be necessary to make new settings in the **Dimension Style Manager** dialog.
6. There are two methods for adding text to a drawing – **Dynamic Text** and **Multiline Text**.

7. When adding text to a drawing, the **Return** key must be used and not the right-hand mouse button.
8. Text styles can be changed during the process of adding text to drawings.
9. AutoCAD 2005 uses two types of text style – **AutoCAD SHX** fonts and **Windows True Type** fonts.
10. Most **True Type** fonts can be in **bold**, **_bold italic_**, _italic_ or regular format. **AutoCAD** fonts can only be added in the single format.
11. When using **Multiline Text**, changes can be made by selection from the popup lists in the **Text Formatting** dialog.
12. To obtain the symbols Ø; ±; °; % use **%%c**; **%%p**; **%%d**; **%%%** respectively before or after the figures of the dimension.
13. Text spelling can be checked with by selecting **Text/Edit** . . . from the **Modify** drop-down menu or by _entering_ **spell** at the command line.

Exercises

1. Open any of the drawings previously saved from working through examples or as answers to exercises and add appropriate dimensions.
2. Construct the drawing Fig. 6.38 but in place of the given dimensions add dimensions showing tolerances of 0.25 above and below.

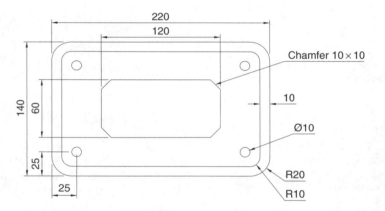

Fig. 6.38 Exercise 2

3. Construct the two polygons in Fig. 6.39 and add all the diagonals. Then set the two osnaps **endpoint** and **intersection** and using the lines as in Fig. 6.39 construct the stars as shown using a polyline of width = 3. Next erase all unwanted lines. Finally dimension the angles labelled **A**, **B**, **C** and **D**.
4. Construct and dimension the drawing in Fig. 6.40.
5. Using the text style **Sans Serif** of height **20** and enclosing the wording within a rectangle of Width = 5 and Fillet = 10, construct the drawing in Fig. 6.41.

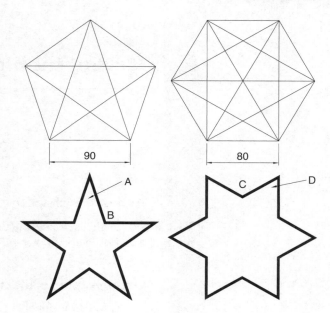

Fig. 6.39 Exercise 3 – Two
polygons

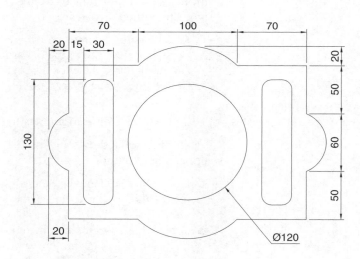

Fig. 6.40 Exercise 4

Fig. 6.41 Exercise 5

AutoCAD 2005

CHAPTER 7

Orthographic and isometric

Fig. 7.1 Example – orthographic projection – the solid being drawn

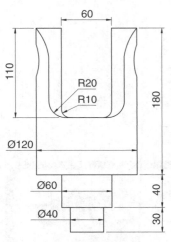

Fig. 7.2 The **front view** of the solid

Aim of this chapter

To introduce methods of constructing drawings of two types – orthographic projection and isometric drawings.

Orthographic projection

Orthographic projection involves viewing an article being described in a technical drawing from different directions – from the front, from a side, from above, from below or from any other viewing position. Orthographic projection often involves:

1. The drawing of details which are hidden, using hidden detail lines;
2. Sectional views in which the article being drawn is imagined as being cut through and the cut surface drawn;
3. Centre lines through arcs, circles, spheres and cylindrical shapes.

An example of an orthographic projection

Taking the solid shown in Fig. 7.1, construct a three-view orthographic projection of the solid:

1. Draw what is seen when the solid is viewed from its left-hand side and regard this as the **front** of the solid. What is drawn will be a **front view** (Fig. 7.2).
2. Draw what is seen when the solid is viewed from the left-hand end of the front view. This produces an **end view**. Fig. 7.3 shows the end view alongside the front view.
3. Draw what is seen when the solid is viewed from above the front view. This produces a **plan**. Fig. 7.4 shows the plan below the front view.
4. Draw centre and hidden detail lines:
    (a) *Click* the arrow to the right of the **Layers** field to show all layers set in the **acadiso.dwt** template on which the drawing has been constructed (Fig. 7.5).
    (b) *Left-click* the **Centre** layer name in the layers list, making it the current layer. All lines will now be drawn as centre lines.
5. In the three-view drawing add centre lines.

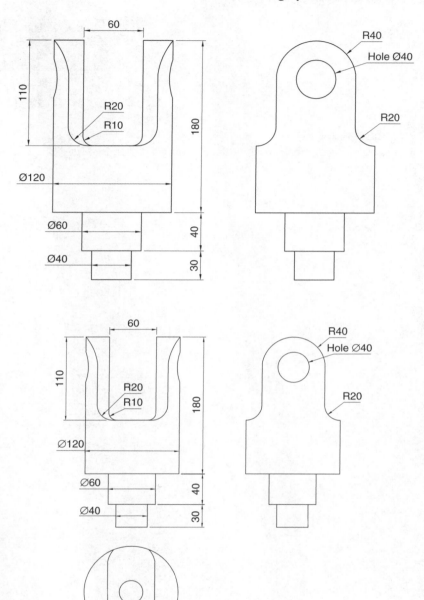

Fig. 7.3 **Front** and **end** views of the solid

Fig. 7.4 **Front** and **end** views, and **plan** of the solid

6. Make the **Hidden** layer the current layer and add hidden detail lines.
7. Make the **Text** layer current and add border lines and a title block.
8. Make the **Dimensions** layer current and add all dimensions.

The completed drawing is shown in Fig. 7.6.

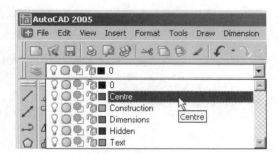

Fig. 7.5 Click the layer name
centre to make it the current layer

Fig. 7.6 The completed working
drawing of the solid

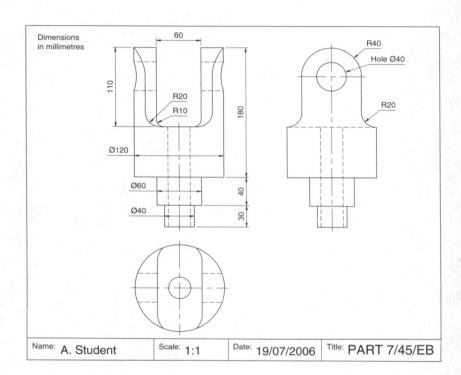

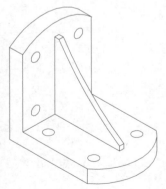

Fig. 7.7 The solid used to
demonstrate first and third
angles of projection

First angle and third angle

There are two types of orthographic projection – **first angle** and **third angle**. Fig. 7.7 is a pictorial drawing of the solid used to demonstrate the two angles. Fig. 7.8 shows a three-view **first angle projection** and Fig. 7.9 shows the same views in **third angle**.

In both angles the viewing is from the same directions. The difference is that the view as seen is placed on the viewing side of the front view in **third angle** and on the opposite side to the viewing in **first angle**.

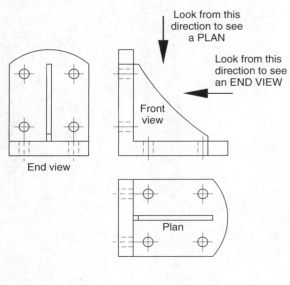

Fig. 7.8 A **first angle** projection

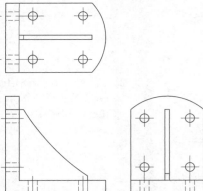

Fig. 7.9 A **third angle** projection

Sectional views

In order to show internal shapes of a solid being drawn in orthographic projection the solid is imagined as being cut along a plane and the cut surface then drawn as seen. Common practice is to **hatch** the areas which then show in the cut surface. Note the section plane line, the section label and the hatching in the sectional view Fig. 7.10.

Adding hatching

To add the hatching as shown in Fig. 7.10:

1. Call the **Hatch** tool – either *left-click* on its tool icon in the **Draw** toolbar (Fig. 7.11) or *enter* **h** at the command line. Note – do not *enter* **hatch** as this gives a different result. The **Boundary Hatch and Fill** dialog (Fig. 7.12) appears.

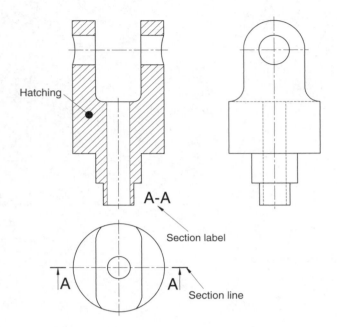

Fig. 7.10 A sectional view

Fig. 7.11 The **Hatch** tool icon
from the **Draw** toolbar

2. *Click* in the **Swatch** field. The **Hatch Pattern Palette** appears. *Left-click* the **ANSI** tab and from the resulting pattern icons *double-click* the **ANSI31** icon. The palette disappears and the **ANSI31** pattern appears in the **Swatch** field.

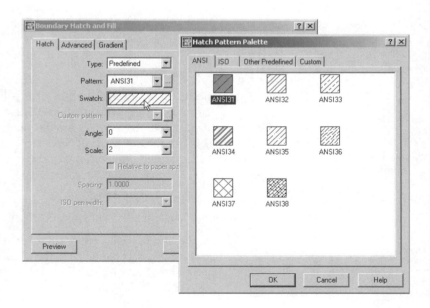

Fig. 7.12 The **Boundary Hatch
and Fill** dialog and the **ANSI
Hatch Pattern Palette**

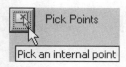

Fig. 7.13 The **Pick Points** button of the **Boundary Hatch and Fill** dialog

3. In the dialog *left-click* the **Pick Points** button (Fig. 7.13). The dialog disappears.

4. In the front view *pick* points as shown in the left-hand drawing of Fig. 7.14. The dialog reappears. *Click* the **Preview** button of the dialog and in the sectional view which reappears, check whether the hatching is satisfactory. In this example it may well be that the **Scale** figure in the dialog needs to be *entered* as **2** in place of the default **1**. Change the figure and **Preview** again. If satisfied *click* the **OK** button of the dialog.

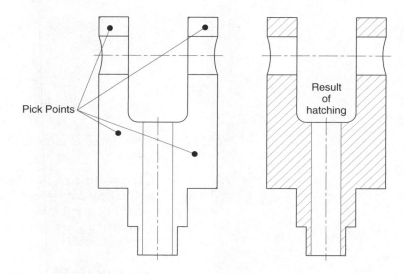

Fig. 7.14 The result of hatching

Isometric drawing

Isometric drawing must not be confused with solid model drawing, examples of which are given in Chapters 13 to 20. Isometric drawing is a 2D method of describing objects in a pictorial form.

Setting the AutoCAD window for isometric drawing

To set the AutoCAD 2005 window for the construction of isometric drawings:

1. At the command line:

Command: *enter* snap
Specify snap spacing or [On/Off/Aspect/Rotate/Style/Type] <5>: s (Style)
Enter snap grid style [Standard/Isometric] <S>: i (Isometric)
Specify vertical spacing <5>: *right-click*
Command:

And the grid dots in the window assume an isometric pattern as shown in Fig. 7.15. Note also the cursor hair lines which are at set in an **Isometric Left** angle.

Fig. 7.15 The AutoCAD grid points set for isometric drawing

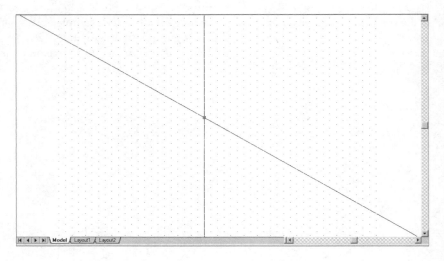

2. There are three isometric angles – **Isometric Top**, **Isometric Right** and **Isometric Left**. These can be set either by pressing the **F5** function key or by pressing the **Ctrl** and **E** keys. Repeated pressing of either of these 'toggles' between the three settings. Fig. 7.16 is an isometric view showing the three isometric planes.

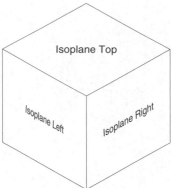

Fig. 7.16 The three isoplanes

The Isometric circle

Circles in an isometric drawing show as ellipses. To add an isometric circle to an isometric drawing, call the **Ellipse** tool. The command line shows:

Command:_ellipse
Specify axis endpoint of ellipse or [Arc/Center/Isocircle]: *enter* **i** (Isocircle) *right-click*
Specify center of isocircle: *pick* or *enter* coordinates
Specify radius of isocircle or [Diameter]: *enter* a number
Command:

And the isocircle appears. Its isoplane position is determined by which of the isoplanes is in operation at the time the isocircle was formed. Fig. 7.17 shows these three isoplanes containing isocircles.

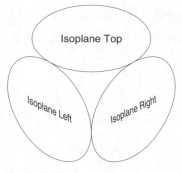

Fig. 7.17 The three isocircles

Examples of isometric drawings

First example – isometric drawing (Fig. 7.20)

1. Work to the shapes and sizes given in the orthographic projection in Fig. 7.18. Set Snap on (press the **F9** function key) and Grid on (**F7**).

2. Set Snap to Isometric and set the isoplane to Isoplane Top using **F5**.
3. With **Line**, construct the outline of the top of the model (Fig. 7.19) working to the dimensions given in Fig. 7.18.

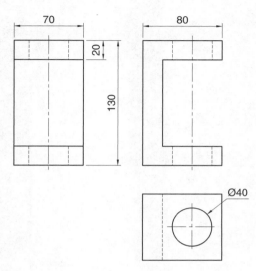

Fig. 7.18 First example –
isometric drawing – the model

4. Call **Ellipse** tool and set to isocircle, and add the isocircle of radius 20 centred in its correct position in the outline of the top (Fig. 7.19).
5. Set the isoplane to Isoplane Right and with **Copy Object** set to **Multiple**, copy the top with its ellipse vertically downwards three times as shown in Fig. 7.19.
6. Add lines as shown in Fig. 7.19.

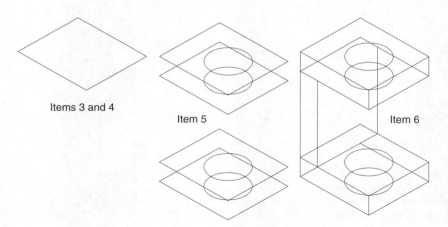

Items 3 and 4

Item 5

Item 6

Fig. 7.19 First example – isometric
drawing – items **3**, **4**, **5** and **6**

7. Finally using **Trim** remove unwanted parts of lines and ellipses to produce Fig. 7.20.

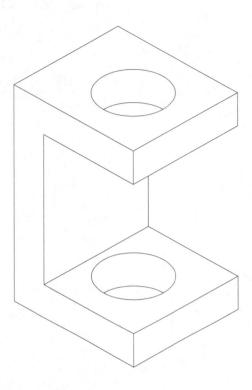

Fig. 7.20 First example – isometric drawing

Second example – isometric drawing (Fig. 7.22)

Fig. 7.21 is an orthographic projection of the model of which the isometric drawing is to be constructed. Fig. 7.22 shows the stages in its construction. The numbers refer to the items in the list below.

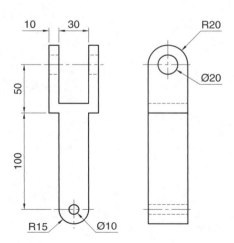

Fig. 7.21 Second example – isometric drawing – orthographic projection of model

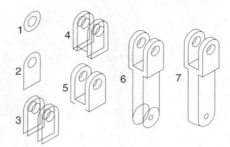

Fig. 7.22 Second example –
isometric drawing – stages in the
construction

1. In **Isoplane Right** construct two isocircles of radii 10 and 20.
2. Add lines as in drawing **2** and trim unwanted parts of isocircle.
3. With **Copy** copy three times as in drawing **3**.
4. Add lines as in drawing **4**.
5. In **Isoplane Left** add lines as in drawing **5**.
6. In **Isoplane Right** add lines and isocircles as in drawing **6**.
7. With **Trim** trim unwanted lines and parts of isocircles to complete the
   isometric drawing – drawing **7**.

Revision notes

1. There are, in the main, two types of orthographic projection – first
   angle and third angle.
2. The number of views included in an orthographic projection depend upon
   the complexity of what is being drawn – a good rule to follow is to attempt
   fully describing the object being drawn in as few views as possible.
3. Sectional views allow parts of an object, which are normally hidden
   from view, to be more fully described in a projection.
4. When a layer is turned **OFF**, all constructions on that layer disappear
   from the screen.
5. If a layer is locked, objects can be added to the layer but no further
   additions or modifications can be made to the layer. If an attempt is
   made to modify an object on a locked layer the command line shows:

Command:_erase
Select objects: *pick* **1 found**
1 was on a locked layer

and the object will not be modified.
6. Frozen layers cannot be selected, but note that layer **0** cannot be frozen.
7. Isometric drawing is a 2D pictorial method of producing illustrations
   showing objects. It is not a 3D method of showing a pictorial view.
8. When drawing ellipses in an isometric drawing, the **Isocircle** prompt
   of the **Ellipse** tool's command line sequence must be used.
9. When constructing an isometric drawing, **Snap** must be set to isometric
   mode before construction can commence.

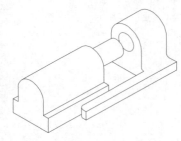

Exercises

Fig. 7.23 is an isometric drawing of a slider fitment on which the three exercises **1**, **2** and **3** are based.

1. Fig. 7.24 is a first angle orthographic projection of part of the fitment shown in the isometric drawing in Fig. 7.23. Construct a three-view third angle orthographic projection of the part.

Fig. 7.23 Exercises 1, 2 and 3 – an isometric drawing of the three parts of the slider on which these exercises are based

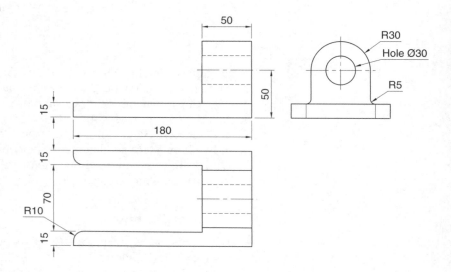

Fig. 7.24 Exercise 1

2. Fig. 7.25 is a first angle orthographic projection of the other part of the fitment shown in Fig. 7.23. Construct a three-view third angle orthographic projection of the part.

3. Construct an isometric drawing of the part shown in Fig. 7.25.

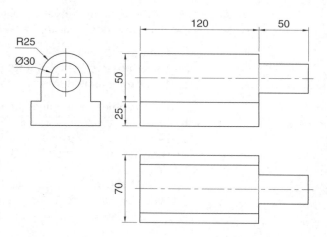

Fig. 7.25 Exercises 2 and 3

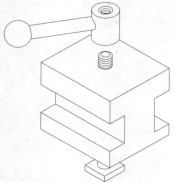

Fig. 7.26 Exercises 4 and 5 – an isometric drawing of the tool holder on which the two exercises are based

4. Construct a three-view orthographic projection, in an angle of your own choice, of the tool holder assembled as shown in the isometric drawing Fig. 7.26. Details are given in Fig. 7.27.

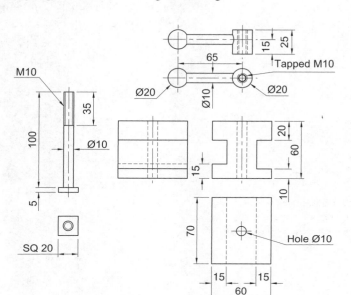

Fig. 7.27 Exercises 4 and 5 – orthographic projections of the three parts of the tool holder

5. Construct an isometric drawing of the body of the tool holder as shown in Figs 7.26 and 7.27.
6. Construct the orthographic projection given in Fig. 7.29.
7. Construct an isometric drawing of the angle plate shown in Figs 7.28 and 7.29.

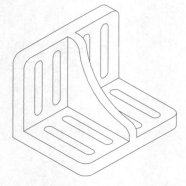

Fig. 7.28 An isometric drawing of the angle plate on which exercises 6 and 7 are based

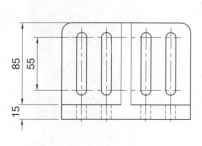
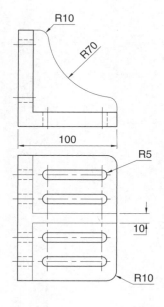

Fig. 7.29 Exercises 6 and 7 – an orthographic projection of the angle plate

8. Construct a third angle projection of the component shown in the iso-metric drawing in Fig. 7.30 and the three-view first angle projection in Fig. 7.31.

9. Construct the isometric drawing shown in Fig. 7.30 working to the dimensions given in Fig. 7.31.

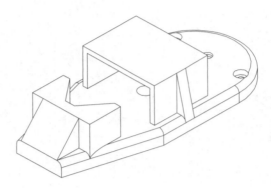

Fig. 7.30 Exercises 8 and 9 – an isometric drawing of the component for the two exercises

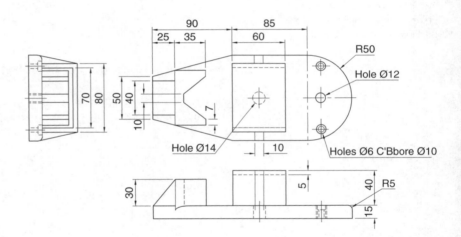

Fig. 7.31 Exercises 8 and 9

CHAPTER 8

Hatching

Aim of this chapter

To describe further examples of the use of hatching in its various forms.

Introduction

In Chapter 7 an example of hatching a sectional view in an orthographic projection was given. Further examples of the use of hatching will be described in this chapter.

There are a large number of hatch patterns available when hatching drawings in AutoCAD 2005. Some examples from the **Hatch Pattern Palette** sub-dialog from the **Other Predefined** set of hatch patterns (Fig. 8.2) are shown in Fig. 8.1.

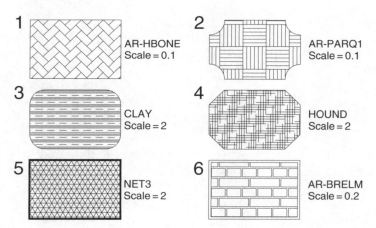

Fig. 8.1 Some hatch patterns from **Predefined** hatch patterns

Other hatch patterns can be selected form the **ISO** or **ANSI** hatch pattern palettes, or the operator can design their own hatch patterns and save them to the **Custom** hatch palette.

First example – hatching a sectional view (Fig. 8.3)

Fig. 8.3 shows a two-view orthographic projection which includes a sectional end view. Note the following in the drawing:

1. The section plane line, consisting of a centre line with its ends marked **A** and an arrow showing the direction of viewing to obtain the sectional view.

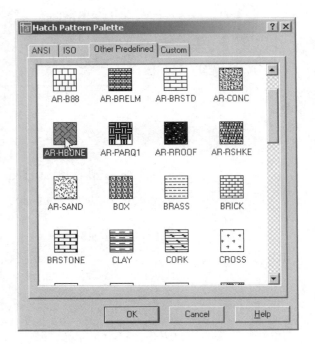

Fig. 8.2 The **Other Predefined** Hatch Pattern Palette

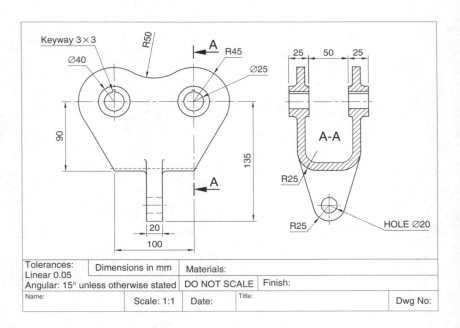

Fig. 8.3 First example – **Hatching**

2. The sectional view labelled with the letters of the section plane line.

3. The cut surfaces of the sectional view hatched with the **ANSI31** hatch pattern which is generally used for the hatching of engineering drawing sections.

Second example – hatching rules (Fig. 8.4)

Fig. 8.4 describes the stages in hatching a sectional end view of a lathe tool holder. Note the following in the section:

1. There are two angles of hatching to differentiate in separate parts of the section.
2. The section follows the general rule that parts such as screws, bolts, nuts, rivets, other cylindrical objects, webs and ribs and other such features are shown within sections as outside views.

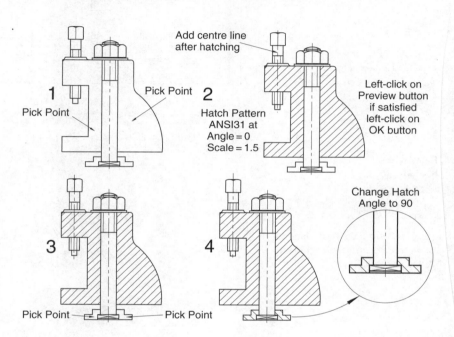

Fig. 8.4 Second example – hatching rules for sections

Third example – Associative hatching (Fig. 8.5)

Fig. 8.5 shows two end views of a house. After constructing the left-hand view, it was found that the upper window had been placed in the wrong position. Using the **Move** tool, the window was moved to a new position. The brick hatching automatically adjusted to the new position. Such **Associative hatching** is only possible if the radio button against **Associative** in the **Composition** area of the **Boundary Hatch and Fill** dialog is **ON**, that is, a dot fills the radio button circle (Fig. 8.6).

Fourth example – Colour gradient hatching (Fig. 8.8)

Fig. 8.8 shows two examples of hatching from the **Gradient** sub-dialog of the **Boundary Hatch and Fill** dialog.

1. Construct two outlines such as those shown in Fig. 8.8.
2. In the **Gradient** sub-dialog (Fig. 8.7) *pick* one of the gradient choices, followed with a *click* on the **Pick Points** button. When the dialog

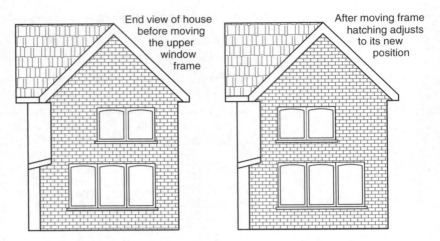

Fig. 8.5 Third example –
Associative hatching

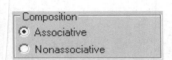

Fig. 8.6 **Associative** hatching set
on in the **Boundary Hatch and
Fill** dialog

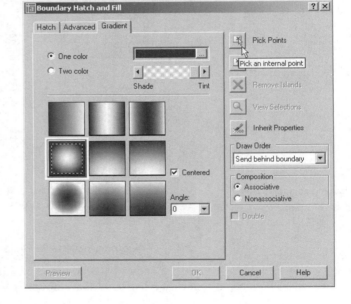

Fig. 8.7 The **Gradient** sub-dialog
of the **Boundary Hatch and Fill**
dialog

disappears, *pick* a single area of one of the drawings, followed by a
click on the dialog's **OK** button when the dialog reappears.

3. Repeat in each of the other areas of the drawing.
4. Change the gradient choice and repeat steps **3** and **4** in the other drawing.

The result is shown in Fig. 8.8.

Notes

1. The colours involved in the gradient hatch can be changed by *clicking*
on the button marked with three dots (. . .) on the right of the colour
field. This brings a **Select Color** dialog on screen, which offers three
choices of sub-dialogs from which to select colours.

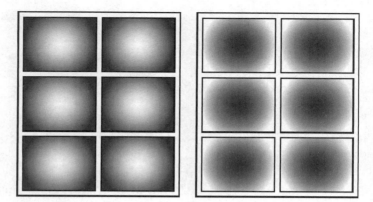

Fig. 8.8 Fourth example –
Gradient hatching

2. The slider marked **Shade** to **Tint** offers changes in shading as the
slider is moved under mouse control.

Fifth example – Advanced hatching (Fig. 8.10)

1. Construct a drawing which includes three outlines as shown in the left-
hand drawing of Fig. 8.10 and copy it twice to produce three identical
drawings.
2. Select the hatch pattern **STARS** at an angle of **0** and scale **1**.
3. *Click* the **Advanced** tab of the dialog, followed by a *click* in the
Normal radio button of the **Island detection style** area (Fig. 8.9).
4. *Pick* a point in the left-hand drawing. The drawing hatches as shown.
5. Repeat in the centre drawing with the radio button of the **Outer** style
set on (dot in button).
6. Repeat in the right-hand drawing with **Ignore** set on.

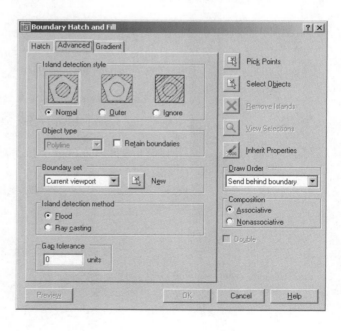

Fig. 8.9 The **Advanced** sub-dialog
of the **Boundary Hatch and Fill**
dialog

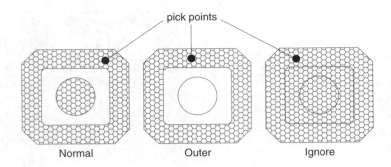

Fig. 8.10 Fifth example –
Advanced hatching

Sixth example – Text in hatching (Fig. 8.11)

1. Construct a pline rectangle using the sizes given in Fig. 8.11.
2. In the **Text Style Manager** dialog, set the text font to **Arial** and its **Height = 25**.
3. Using the **Dtext** tool *enter* the text as shown central to the rectangle.
4. Hatch the area using the **HONEY** hatch pattern set to an angle of **0** and scale of **1**.

The result is shown in Fig. 8.11.

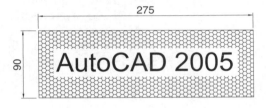

Fig. 8.11 Sixth example – **Text in hatching**

Note

Text will be entered with a surrounding boundary area free from hatching, providing the **Advanced Normal** radio button is set on.

Seventh example – Advanced hatching (Fig. 8.17)

1. Open the **Layer Properties Manager** with a *click* on its icon in the **Layer** toolbar.
2. Make layers as shown in Fig. 8.12.

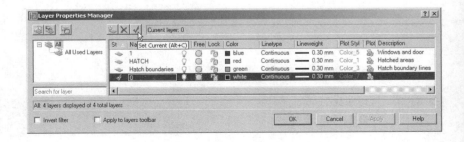

Fig. 8.12 Seventh example – the layers setup for the advanced hatch example

3. With the layer **0** current, construct the outline as given in Fig. 8.13.

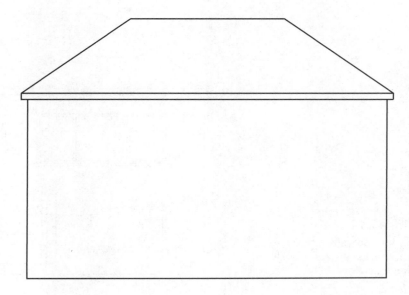

Fig. 8.13 Seventh example –
construction on layer **0**

4. With layer **1** current, construct the door and windows (Fig. 8.14).

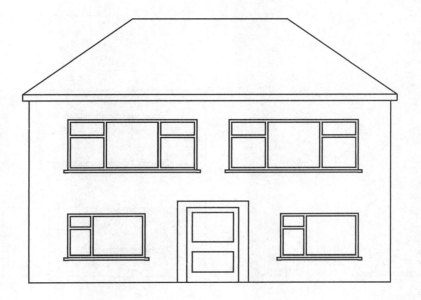

Fig. 8.14 Seventh example –
construction on layer **1**

5. With layer **Hatch boundaries** current, construct the lines as shown in Fig. 8.15.

Fig. 8.15 Seventh example – construction on layer **Hatch boundaries**

6. Make the layer **HATCH** current and add hatching to the areas shown in Fig. 8.16 using the hatch patterns **ANGLE** at scale **2** for the roof and **BRICK** at a scale of **0.75** for the wall.

Fig. 8.16 Seventh example – construction on layer **HATCH**

7. Finally turn the layer **Hatch boundaries** off. The result is given in Fig. 8.17.

Fig. 8.17 Seventh example – the finished drawing

Revision notes

1. A large variety of hatch patterns are available when working with AutoCAD 2005.
2. In sectional views in engineering drawings it is usual to show items such as bolts, screws, other cylindrical objects, webs and ribs as outside views.
3. When associative hatching is set on, if an object is moved within a hatched area, the hatching accomodates to fit around the moved object.
4. Colour gradient hatching is available in AutoCAD 2005.
5. When hatching takes place around text, a space around the text will be free from hatching.

Exercises

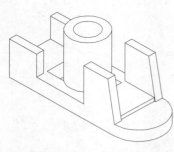

Fig. 8.18 Exercise 1 – a pictorial view

1. Fig. 8.18 shows a pictorial drawing of the component shown in the three-view orthographic projection in Fig. 8.19. Construct the three views with the front view as a sectional view based on the section plane **A-A**.

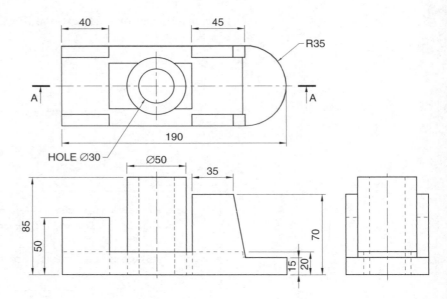

Fig. 8.19 Exercise 1

2. Construct the three-view orthographic projection in Fig. 8.20 to the given dimensions with the front view as the sectional view **A-A**.

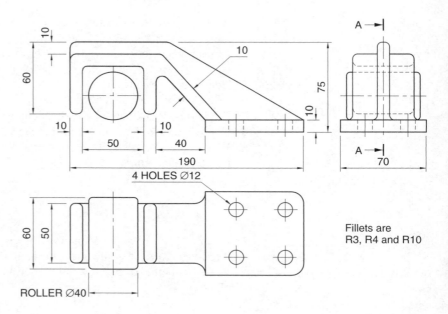

Fig. 8.20 Exercise 2

3. Construct the drawing **Stage 5** following the descriptions of stages given in Fig. 8.21.

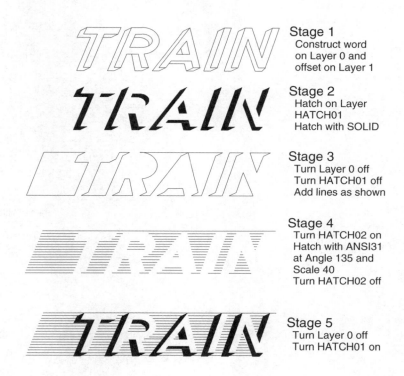

Stage 1
Construct word
on Layer 0 and
offset on Layer 1

Stage 2
Hatch on Layer
HATCH01
Hatch with SOLID

Stage 3
Turn Layer 0 off
Turn HATCH01 off
Add lines as shown

Stage 4
Turn HATCH02 on
Hatch with ANSI31
at Angle 135 and
Scale 40
Turn HATCH02 off

Stage 5
Turn Layer 0 off
Turn HATCH01 on

Fig. 8.21 Exercise 3

4. Fig. 8.22 is a front view of a car with parts hatched. Construct a similar drawing of any make of car, using hatching to emphasise the shape.

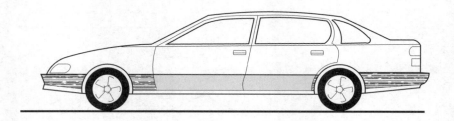

Fig. 8.22 Exercise 4

5. Working to the notes given with the drawing in Fig. 8.23, construct the end view of a house as shown. Use your own discretion about sizes for the parts of the drawing.

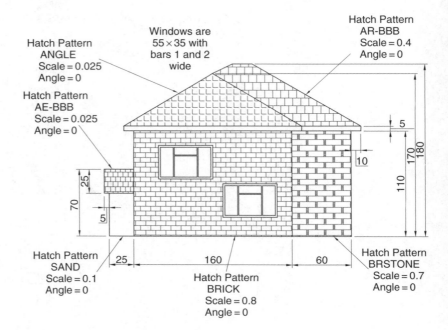

Fig. 8.23 Exercise 5

6. Working to dimensions of your own choice, construct the three-view projection of a two-storey house as shown in Fig. 8.24.

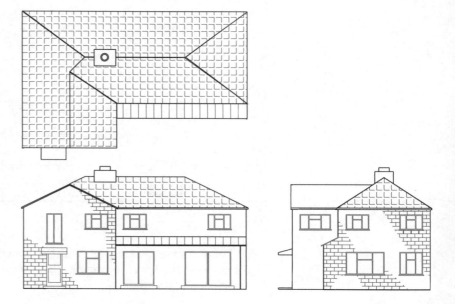

Fig. 8.24 Exercise 6

7. Construct Fig. 8.25 as follows:
   (a) On layer **Text**, construct a circle of radius **90**.
   (b) Make layer **0** current.
   (c) Construct the small drawing to the details as shown and save as a block with a block name **shape** (see Chapter 9).
   (d) Call the **Divide** tool by *entering* **div** at the command line:

   Command: *enter* **div** *right-click*
   Select object to divide: *pick* the circle
   Enter number of segments or [Block]: *enter* **b** *right-click*
   Enter name of block to insert: *enter* **shape** *right-click*
   Align block with object? [Yes/No] <Y>: *right-click*
   Enter the number of segments: *enter* **20** *right-click*
   Command:

   (e) Turn the layer **Text** off.

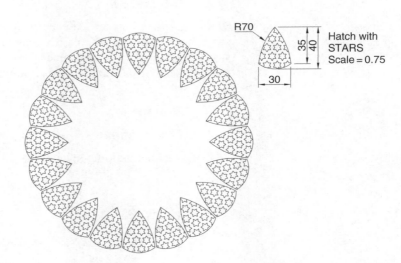

Fig. 8.25 Exercise 7

CHAPTER 9

Blocks and Inserts

Aims of this chapter

1. To describe the construction of **blocks** and **wblocks** (written blocks).
2. To introduce the insertion of blocks and wblocks into other drawings.
3. To introduce the use of the **DesignCenter** palette.
4. To explain the use of **Explode** and **Purge** tools.

Introduction

Blocks are drawings which can be inserted into other drawings. Blocks are contained in the data of the drawing in which they have been constructed. Wblocks (written blocks) are saved as drawings in their own right, but can be inserted into other drawings if required.

Blocks

First example – Blocks (Fig. 9.3)

1. Construct the building symbols as shown in Fig. 9.1 to a scale of 1:50.

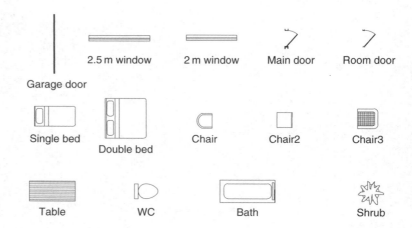

Fig. 9.1 First example –
Blocks – symbols to be saved
as blocks

2. *Left-click* the **Make Block** tool (Fig. 9.2). The **Block Definition** dialog (Fig. 9.3) appears. To make a block of the **Double bed** symbol drawing:

Fig. 9.2 The **Make Block** tool icon from the **Draw** toolbar

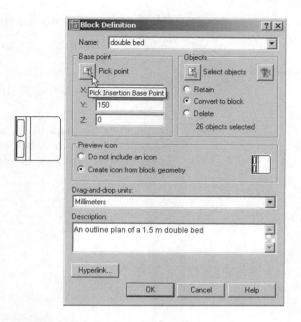

Fig. 9.3 The **Block Definition** dialog with entries for the **double bed**

(a) *Enter* **double bed** in the **Name** field.

(b) *Click* the **Select Objects** button. The dialog disappears. Window the drawing of the double bed. The dialog reappears. Note the icon of the double bed in the **Preview icon** area of the dialog.

(c) *Click* the **Pick Point** button. The dialog disappears. *Click* a point on the double bed drawing to determine its **insertion point**. The dialog reappears. Note the coordinates of the insertion point, which appear in the **Base point** area of the dialog.

(d) If thought necessary *enter* a description in the **Description** field of the dialog.

The drawing is now saved as a **block** in the drawing.

3. Repeat steps **1** and **2** to make blocks of all the other symbols in the drawing.

4. Open the **Block Definition** dialog again and *click* the arrow on the right of the **Name** field. The blocks saved in the drawing appear in a popup list (Fig. 9.4).

Fig. 9.4 The popup list in the **Name** field showing all blocks saved in the drawing

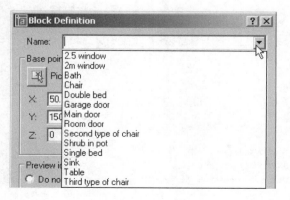

Inserting blocks into a drawing

There are two methods by which symbols saved as blocks can be inserted into a drawing.

Example – first method of inserting blocks

Ensuring that all the symbols saved as blocks using the **Make Block** tool are saved in the data of the drawing in which the symbols were constructed, erase the drawings of the symbols and in their place construct the outline of the plan of a bungalow to a scale of 1:50 (Fig. 9.5). Then:

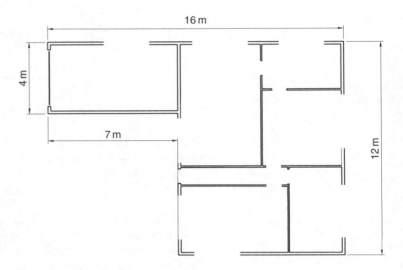

Fig. 9.5 First example – inserting blocks. Outline plan

Fig. 9.6 The **Insert Block** tool icon in the **Draw** toolbar

1. *Left-click* the **Insert Block** tool icon in the **Draw** toolbar (Fig. 9.6). The **Insert** dialog appears on screen (Fig. 9.7). From the **Name** popup list select the name of the block which is to be inserted – in this example the **2.5 window**.

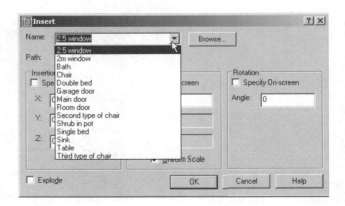

Fig. 9.7 The **Insert** dialog with its **Name** popup list displaying the names of all blocks in the drawing

2. Make sure the check box against **Explode** is off (no tick in box). *Click* the dialog's **OK** button, the dialog disappears. The symbol drawing appears with its insertion point at the intersection of the cursor hairs ready to be *dragged* into its position in the plan drawing.
3. Once all the block drawings are placed, their positions can be adjusted. Blocks are single objects and can thus be dragged into new positions as required under mouse control. Their angle of position can be amended at the command line which shows:

Command:
INSERT
Specify insertion point or [Scale.X/Y/Z/Rotate/PScale/PX/PY/PZ/ PROTATE]: *enter* **r** (Rotate) *right-click*
Specify insertion angle: *enter* **180** *right-click*
Specify insertion point: *pick*
Command:

Selection from these prompts allows scaling, stretching along any axis, previewing, etc. as the block is inserted.
4. Insert all necessary blocks and add other details as required to the plan outline drawing. The result is given in Fig. 9.8.

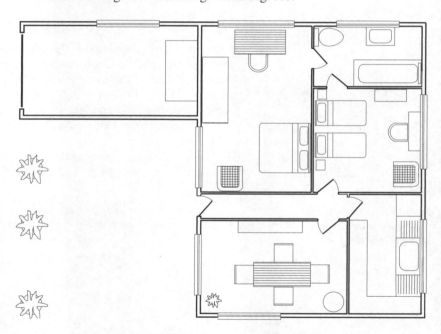

Fig. 9.8 Example – first method of inserting blocks

Example – second method of inserting blocks

1. Save the drawing which includes all the blocks to a suitable file name (building_symbols.dwg). Remember this drawing includes data of the blocks in its file.
2. *Left-click* the **DesignCenter** icon in the **Standard** toolbar (Fig. 9.9). The **DesignCenter** palette appears on screen (Fig. 9.10).

DesignCenter (Ctrl+2)

Fig. 9.9 The **DesignCenter** icon
in the **Standard** toolbar

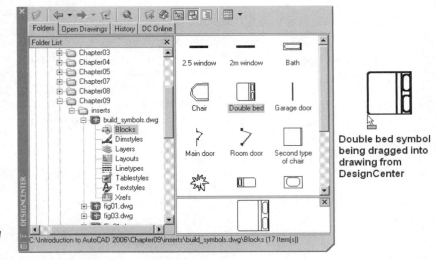

Double bed symbol
being dragged into
drawing from
DesignCenter

Fig. 9.10 The **DesignCenter**
with the double bed block *dragged*
on screen

3. With the outline plan (Fig. 9.5) on screen the symbols can all be
   dragged into position from the **DesignCenter**.

Notes about DesignCenter palette

1. As with other palettes, the **DesignCenter** palette can be re-sized by
   dragging the palette to a new size from its edges or corners.
2. *Clicks* on one of the three icons at the top-right corner of the palette
   (Fig. 9.11) have the following results:
   Tree View Toggle – changes from showing two areas – a **Folder List**
   and icons of the blocks within a file – to a single area showing the
   block icons.
   Preview – a *click* on the icon opens a small area at the base of the palette
   showing an enlarged view of the selected block icon (Fig. 9.12).

Fig. 9.11 The icons at the
top-right corner of the
DesignCenter palette

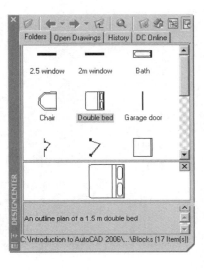

Fig. 9.12 The results of *clicks* on
Tree View Toggle, **Preview**
and **Description**

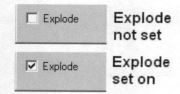

Fig. 9.13 The **Explode** check box
in the **Insert dialog**

Description – a *click* on the icon opens another small area with a description of the block.

The Explode and Purge tools

A block is a single object no matter from how many objects it was originally constructed. This enables a block to be *dragged* about the drawing area as a single object.

A check box in the bottom left-hand corner of the **Insert** dialog is labelled **Explode** (Fig. 9.13). If the check box is ticked, **Explode** will be set on and when a block is inserted it will be exploded into the objects from which it was constructed.

Another way of exploding a block would be to use the **Explode** tool from the **Modify** toolbar (Fig. 9.14). A *click* on the icon or *entering* **x** at the command line brings prompts into the command line:

Fig. 9.14 The **Explode** tool in the
Modify toolbar

Command:_explode
Select objects: *pick* a block on screen **1 found.**
Select objects: *right-click*
Command:

And the *picked* object is exploded into its original objects.

The Purge tool

The **Purge** tool can be called by *entering* **pu** at the command line or from **Drawing Utilities** in the **File** drop-down menu (Fig. 9.15). When the tool is called the **Purge** dialog appears on screen (Fig. 9.16).

Fig. 9.15 Calling **Purge** from the
File drop-down menu

The **Purge** tool can be used to remove the data of blocks within a drawing thus saving file space when a drawing which includes blocks is saved to file.

To use the tool, in its dialog *click* the **Purge** button and a sub-dialog appears naming a block to be purged. A *click* on the **Yes** button clears the data of the block from the drawing. Continue until all blocks that are to be purged are removed.

Take the drawing in Fig. 9.8 (page 127) as an example. If all the blocks are purged from the drawing, the file will be reduced from **145** KB to **67** KB when the drawing is saved to disk.

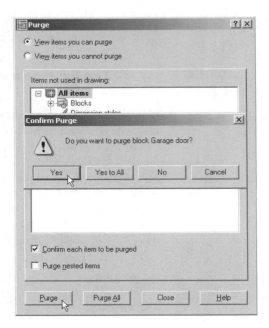

Fig. 9.16 The **Purge** dialog

Example using the DesignCenter (Fig. 9.19)

1. Construct the set of electric/electronic circuit symbols shown in Fig. 9.17 and make a series of blocks from each of the symbols.

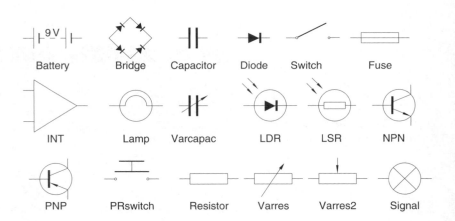

Fig. 9.17 Example using the DesignCenter – a set of electric/electronic symbols

2. Save the drawing to a file **Electrics.dwg**.
3. Open the **acadiso.dwt** template. Open the **DesignCenter** with a *click* on its icon in the **Standard** toolbar.
4. From the **Folder list** select the file **electrics.dwg** and *click* on **Blocks** under its file name. Then *drag* symbols from the symbol icons from the **DesignCenter** into the drawing area as shown in Fig. 9.18. Ensure

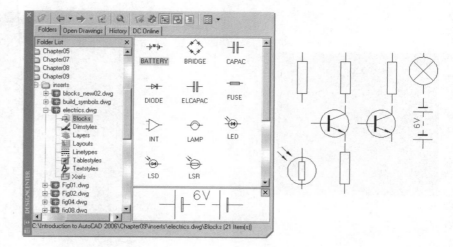

Fig. 9.18 Example using the DesignCenter

they are placed in appropriate positions in relation to each other to form a circuit. If necessary either **Move** or/and **Rotate** the symbols into correct positions.

5. Close the **DesignCenter** palette with a *click* on the **x** in the top left-hand corner.
6. Complete the circuit drawing as shown in Fig. 9.19.

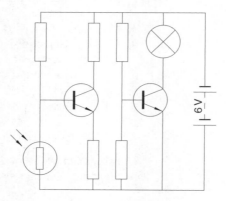

Fig. 9.19 Example using the DesignCenter – the completed circuit

Note

Fig. 9.19 does not represent an authentic electronics circuit.

Wblocks

Wblocks or written blocks are saved as drawing files in their own right and are not part of the drawing in which they have been saved.

Example – wblock (Fig. 9.20)

1. Construct a light emitting diode (**LED**) symbol and *enter* **w** at the command line. The **Write Block** dialog appears (Fig. 9.20).

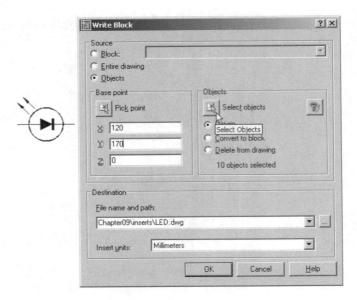

Fig. 9.20 Example – **Wblock**

2. *Click* the button marked with three dots (. . .) to the right of the **File name and path** field and from the **Browse for Drawing File** dialog which comes to screen select an appropriate directory. The directory name appears in the **File name and path** field. Add **LED.dwg** at the end of the name.

3. Make sure the **Insert units** is set to **Millimetres** in its popup list.

4. *Click* the **Select objects** button, window the symbol drawing and when the dialog reappears, *click* the **Pick point** button, followed by selecting the left-hand end of the symbol.

5. Finally *click* the **OK** button of the dialog and the symbol is saved in its selected directory as a drawing file **LED.dwg** in its own right.

Notes on the DesignCenter

Drawings can be inserted into the AutoCAD window from the **DesignCenter** by dragging the icon representing the drawing into the window (Fig. 9.21).

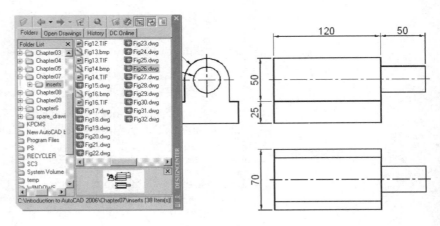

Fig. 9.21 An example of a drawing *dragged* from the **DesignCenter**

When such a drawing is *dragged* into the AutoCAD window, the command line shows a sequence such as:

**Command:_INSERT Enter block name or [?] <Fig26>: "Chapter07\
inserts\Fig26.dwg"**
Specify insertion point or [prompts]: *pick*
Enter X scale factor <1>: *right-click*
Enter Y scale factor <use X scale factor>: *right-click*
Specify rotation angle <0>: *right-click*
Command:

Revision notes

1. Blocks become part of the drawing file in which they were constructed.
2. Wblocks become drawing files in their own right.
3. Drawings or parts of drawings can be inserted in other drawings with the **Insert Block** tool.
4. Inserted blocks or drawings are single objects unless either the **Explode** check box of the **Insert** dialog is checked, or the block or drawing is exploded with the **Explode** tool.
5. Drawings can be inserted into the AutoCAD drawing area using the **DesignCenter**.
6. Blocks within drawings can be inserted into drawings from the **DesignCenter**.

Exercises

1. Construct the building symbols in Fig. 9.22 in a drawing saved as **symbols.dwg**. Then using the **DesignCenter** construct a building drawing of the first floor of the house you are living in, making use of the symbols. Do not bother too much about dimensions because this exercise is designed to practise using the idea of making blocks and using the **DesignCenter**.

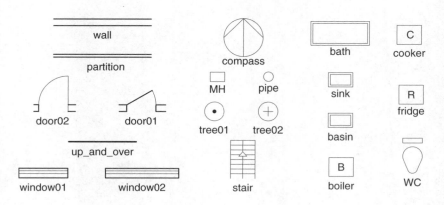

Fig. 9.22 Exercise 1

2. Construct drawings of the electric/electronics symbols in Fig. 9.17 (page 130) and save them as blocks in a drawing file **electronics.dwg**.

3. Construct the electronics circuit given in Fig. 9.23 from the file **Electrics.dwg** using the **DesignCenter**.

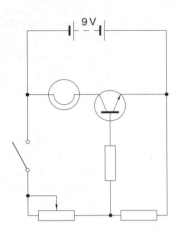

Fig. 9.23 Exercise 3

4. Construct the electronics circuit given in Fig. 9.24 from the file **electronics.dwg** using the **DesignCenter**.

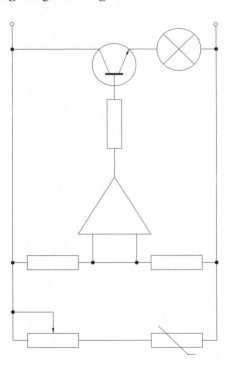

Fig. 9.24 Exercise 4

CHAPTER 10

Other types of file format

Aims of this chapter

1. To introduce Object Linking and Embedding (**OLE**) and its uses.
2. To introduce the use of Encapsulated Postscript (**EPS**) files.
3. To introduce the use of Data Exchange Format (**DXF**) files.
4. To introduce raster files.
5. To introduce **Xrefs**.

Object Linking and Embedding

First example – copying and pasting (Fig. 10.3)

1. Open any drawing in the AutoCAD 2005 window (Fig. 10.1).
2. *Left-click* **Copy Link** in the **Edit** drop-down menu (Fig. 10.1).

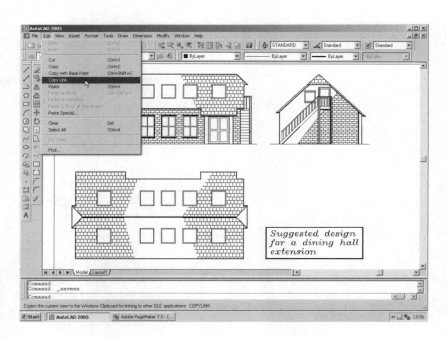

Fig. 10.1 A drawing in the AutoCAD 2005 window showing **Copy Link** selected from the **Edit** drop-down menu

3. *Click* the AutoCAD 2005 **Minimize** button and open the **Clipboard** viewer. The copied drawing appears in the clipboard (Fig. 10.2).

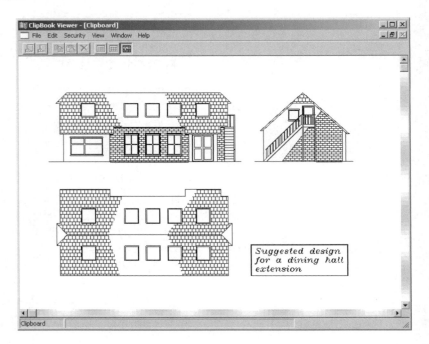

Fig. 10.2 The drawing from AutoCAD copied to the **Clipboard**

4. Open **Microsoft Word** and *click* on **Paste** in the **Edit** drop-down menu (Fig. 10.3). The drawing from the **Clipboard** appears in the **Microsoft Word** document (Fig. 10.3).
5. Add text as required.

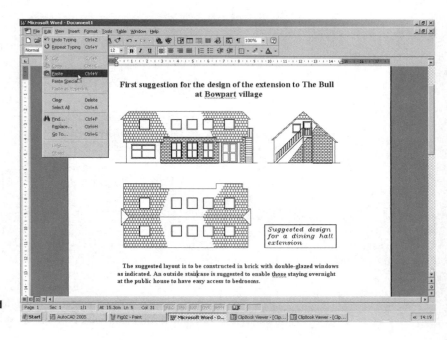

Fig. 10.3 Example – **copying and pasting**

Notes

1. It is not common practice to have a **Clipboard** window showing on screen. The Clipboard usually works in the background and thus it is not normal practice for it to be seen. It is shown opened here to display its use in acting as an agent for transposing drawings etc. from one application to another.
2. Similar results can be obtained using the **Copy** and **Copy with Base Point** tools from the **Edit** drop-down menu of AutoCAD 2005.
3. The drawing could also be pasted back into the AutoCAD window – not that there would be much point in doing so, but anything in the **Clipboard** window can be pasted into other applications.

Second example – EPS file (Fig. 10.5)

1. With the same drawing on screen *click* on **Export** . . . in the **File** drop-down menu. The **Export Data** dialog appears (Fig. 10.4). *Pick* **Encapsulated PS** (**.eps*) from the **Files of type** popup list, then *enter* a suitable file name (**building.eps**) in the **File name** field and *click* the **Save** button.

Fig. 10.4 The **Export Data** dialog of AutoCAD 2005

2. Open a desktop publishing application. That shown in Fig. 10.5 is **PageMaker**.
3. From the **File** drop-down menu *click* **Place** . . . A dialog appears listing files which can be placed in a PageMaker document. Among the files named will be **building.eps**. *Double-click* that file name and an icon appears, the placing of which determines the position of the *.eps file drawing in the PageMaker document (Fig. 10.5).

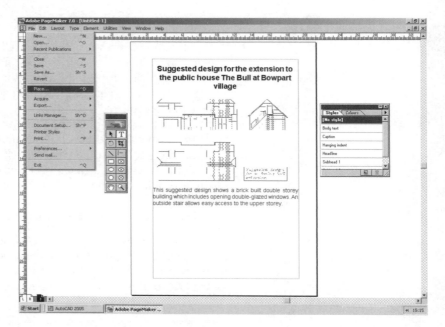

Fig. 10.5 An EPS file placed in position in a PageMaker document

4. Add text as required.

5. Save the PageMaker document to a suitable file name.

6. Go back to the AutoCAD drawing and delete the title.

7. Make a new *.eps file with the same file name (**building.eps**).

8. Go back into **PageMaker** and *click* **Links Manager** . . . in the **File** drop-down menu. The **Links Manager** dialog appears (Fig. 10.6). Against the name of the **building.eps** file name is a dash and a note at the bottom of the dialog explaining that changes have taken place in the drawing from which the *.eps had been derived. *Click the* **Update** button and when the document reappears the drawing in PageMaker no longer includes the erased title.

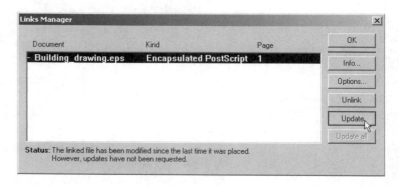

Fig. 10.6 The **Links Manager** dialog of PageMaker

Notes

1. This is **Object Linking and Embedding**. Changes in the AutoCAD drawing saved as an *.eps file are linked to the drawing embedded in another application document, so changes made in the AutoCAD drawing are reflected in the PageMaker document.

2. There is actually no need to use the **Links Manager** because if the file from PageMaker is saved with the old *.eps file in place, when it is reopened the file will have changed to the redrawn AutoCAD drawing, without the erased title.

DXF (Data Exchange Format) files

The *.**DXF** format was originated by Autodesk (publishers of AutoCAD), but is now in general use in most **CAD** software. A drawing saved to a *.**dxf** format file can be opened in most other CAD software applications. This file format is of great value when drawings are being exchanged between operators using different CAD applications.

Example – DXF file (Fig. 10.8)

1. Open a drawing in AutoCAD. This example is shown in Fig. 10.7.

Fig. 10.7 Example – **DXF file**.
Drawing to be saved as a dxf file

2. *Click* on **Save As** . . . in the **File** drop-down menu and in the **Save Drawing As** dialog which appears, *click* **AutoCAD 2005 DXF [*.dxf]**.
3. *Enter* a suitable file name. In this example this is **Fig07**. The extension **.dxf** is automatically included when the **Save** button of the dialog is *clicked* (Fig. 10.8).
4. The **DXF** file can now be opened in the majority of CAD applications and then saved to the drawing file format of the CAD in use.

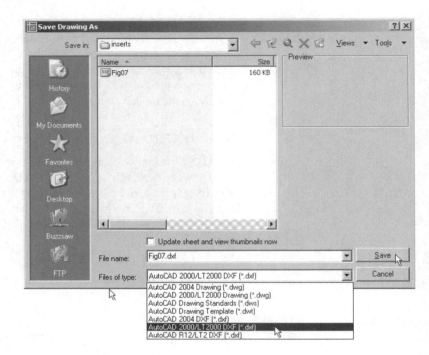

Fig. 10.8 The **Save Drawing As** dialog set up to save drawings in **DXF** format

Note

To open a **DXF** file in AutoCAD 2005, select **Open** . . . from the **File** drop-down menu and select **AutoCAD 2005 DXF** [***.dxf**] from the popup list from the **Files of type** field.

Raster images

A variety of raster files can be placed into AutoCAD 2005 drawings from the **Select Image File** dialog brought to screen with a *click* on **Raster Image** . . . in the **Insert** drop-down menu. In this example the selected raster file is a bitmap (extension ***.bmp**) of a rendered 3D model drawing constructed to the views in an assembly drawing of a lathe tool post (see Chapter 16 for rendering 3D models).

Example – placing a raster file in a drawing (Fig. 10.12)

1. *Click* **Raster Image** . . . from the **Insert** drop-down menu (Fig. 10.9). The **Select Image File** dialog appears (Fig. 10.10). *Click* the file name of the image to be inserted – in this example **rendering01.bmp**. A preview appears in the **Preview** area of the dialog.
2. *Click* the **Open** button of the dialog. The **Image** dialog appears (Fig. 10.11).
3. In the **Scale** field *enter* a suitable scale figure. The size of the image that will appear in the AutoCAD window can be seen with a *click* on

Fig. 10.9 Selecting **Raster Image** . . . from the **Insert** drop-down menu

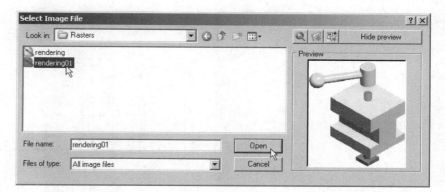

Fig. 10.10 The **Select Image File** dialog

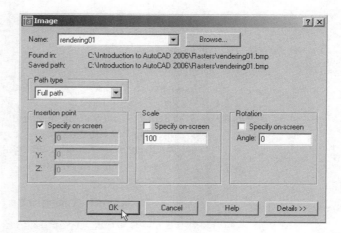

Fig. 10.11 The **Image** dialog

the **Details** button which brings down an extension of the dialog which shows details about the resolution and size of the image.

4. *Click* the **OK** button, the command line shows:

> **Command:_imageattach**
> **Specify insertion point <0,0>:** an outline of the image attached to the intersection of the cursor cross hairs appears *pick* a suitable point on screen
> **Command:**

And the raster image appears at the *picked* point (Fig. 10.12).

Note

As can be seen from the **Insert** drop-down menu (Fig. 10.9), a variety of different types of raster and other images can be inserted into an Auto-CAD drawing. Some examples are:

Blocks – see Chapter 9.
External References (Xrefs) – see later in this chapter.
Field name – a *click* on the name brings up the **Field** dialog. Practise inserting various categories of field names from the dialog.

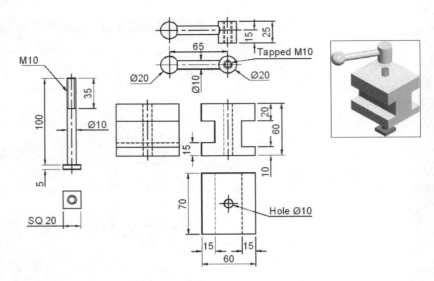

Fig. 10.12 Example – placing a
raster file in a drawing

Layout – a wizard appears allowing new layouts to be created and saved for new templates.

3D Studio – allows the insertion of images, constructed in the Autodesk software **3D Studio**, from files with the format *.3ds*.

OLE Objects – allows raster images to be placed as OLE images from a variety of other applications.

External References (Xrefs)

If a drawing is inserted into another drawing as an external reference, any changes made in the original Xref drawing are automatically reflected in the drawing into which the Xref has been inserted.

Example – Xrefs (Fig. 10.19)

1. Construct the three-view orthographic drawing in Fig. 10.15. Dimensions of this drawing will be found on page 230. Save the drawing to a suitable file name.
2. As a separate drawing construct Fig. 10.16. Save it as a wblock with the name of **spindle.dwg** and with a base insertion point at one end of its centre line.
3. In the **Insert** drop-down menu *click* **Xref Manager** . . . The **Xref Manager** dialog appears (Fig. 10.13).
4. *Click* its **Attach** button. The **Select Reference File** dialog appears. Select the **spindle** drawing from the file list. Its name appears in **Name** field of another dialog **External Reference** (Fig. 10.14). *Click* its **OK** button.

Fig. 10.13 The **Xref Manager** dialog

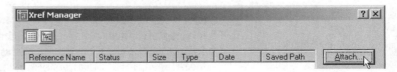

Fig. 10.14 The **External Reference** dialog

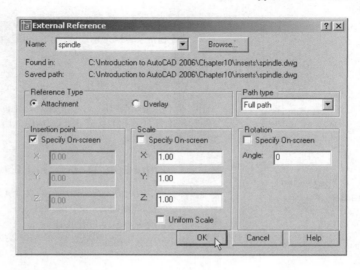

Fig. 10.15 Example – **Xrefs** – original drawing

Fig. 10.16 The **spindle.dwg** drawing

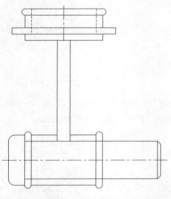

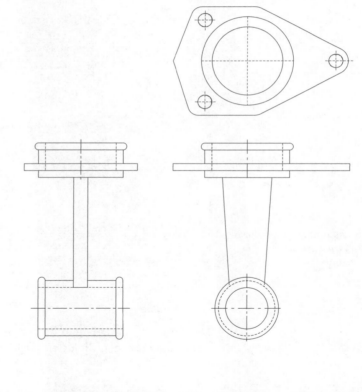

Fig. 10.17 The spindle in place in the original drawing

5. The spindle drawing appears on screen ready to be *dragged* into position. Place it in position as indicated in Fig. 10.17.
6. Save the drawing with its Xref to its original file name.
7. Open the drawing **spindle.dwg** and make changes as shown in Fig. 10.18.

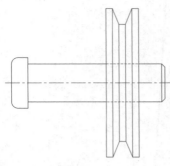

Fig. 10.18 The revised
spindle.dwg drawing

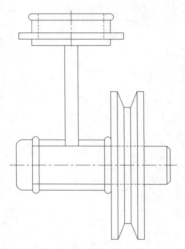

Fig. 10.19 Example – **Xrefs**

Fig. 10.20 Four drawings in the
**Multiple Document
Environment**

8. Now reopen the original drawing. The **Xref** within the drawing has changed in accordance with the alterations to the spindle drawing (Fig. 10.19).

Note

In this example to ensure accuracy of drawing the **Xref** will need to be exploded and parts of the spindle changed to hidden detail lines.

Multiple Document Environment (MDE)

1. Open several drawings in AutoCAD – in this example four separate drawings have been opened.
2. In the **Window** drop-down menu, *click* **Tile Horizontally**. The four drawings rearrange as shown in Fig. 10.20.

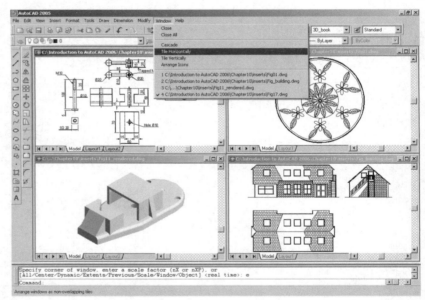

Note

The names of the drawings appear at the base of the **Window** drop-down menu, showing their directories, file names and file name extensions.

Revision notes

1. When an *.eps file is inserted from AutoCAD 2005 into a document, subsequent changes in the original drawing are reflected in the inserted *.eps file in the document.

2. The **Edit** tools **Copy with Base Point**, **Copy** and **Copy Link** enable objects from AutoCAD 2005 to be copied for **Pasting** onto other applications.

3. Objects can be copied from other applications to be pasted into the AutoCAD 2005 window.

4. Drawings saved in AutoCAD as **DXF** (*.**dxf**) files can be opened in other CAD applications.

5. Similarly drawings saved in other CAD applications as *.**dxf** files can be opened in AutoCAD 2005.

6. **Raster** files of the format types *.**bmp**, *.**jpg**, *.**pcx**, *.**tga**, *.**tif**, among other raster type file objects, can be inserted into AutoCAD 2005 drawings.

7. Drawings saved to the Encapsulated Postscript (*.**eps**) file format can be inserted into documents of other applications.

8. Changes made in a drawing saved as an *.**eps** file will be reflected in the drawing inserted as an *.**eps** file in another application.

9. A number of drawings can be opened in the AutoCAD 2005 window.

Exercises

1. Fig. 10.21 shows a pattern formed by inserting an **Xref** and then either copying or arraying the **Xref**.

Fig. 10.21 Exercise 1 – original pattern

The hatched parts of the **Xref** drawing were then changed using a different hatch pattern. The result of the change in the **Xref** is shown in Fig. 10.22.

Construct a similar **Xref** drawing, insert an **Xref**, array or copy to form the pattern, then change the hatching, save the **Xref** drawing and note the results.

Fig. 10.22 Exercise 1

Fig. 10.23 Exercise 2 – a rendering of the holders and roller

2. Fig. 10.23 is a rendering of a roller between two end holders. Fig. 10.24 gives details of the end holders and the roller in orthographic projections.

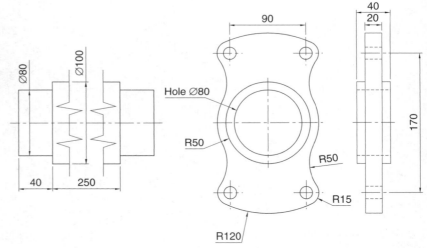

Fig. 10.24 Exercise 2 – details of the parts of the roller and holders

Construct a full size front view of the roller and save to a file name **roller.dwg**. Then as a separate drawing construct a front view of the two end holders in their correct positions to receive the roller and save to the file name **assembly.dwg**.

Insert the roller drawing into the assembly drawing as an **Xref**.

Open the **roller.dwg** and change its outline as shown in Fig. 10.25. Save the drawing. Open the **assembly.dwg** and note the change in the inserted Xref.

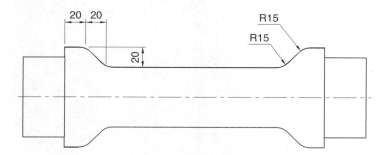

Fig. 10.25 The amended **Xref** darwing

3. *Click* **Raster Image** . . . in the **Insert** drop-down menu and insert a **JPEG** image (*.jpg file) of a photograph into the AutoCAD 2005 window. An example is given in Fig. 10.26.

Fig. 10.26 Exercise 3 – example

4. Using **Copy** from the **Insert** drop-down menu, copy a drawing from AutoCAD 2005 into a Microsoft Word document. An example is given in Fig. 10.27. Add some appropriate text.

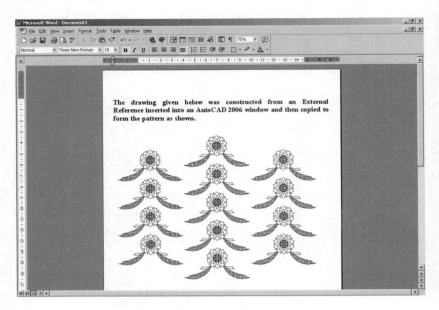

Fig. 10.27 Exercise 4 – an example

5. The plan in Figs 10.1, 10.2 and 10.3 is incorrect in that some details have been missed from the drawing. Can you identify the error?

CHAPTER 11

Sheet sets

Aims of this chapter

1. To introduce sheet sets.
2. To give an example of a sheet set based on the design of a two-storey house.

Sheet sets

When anything is to be manufactured or constructed, whether it be a building, an engineering design, an electronics device or any other form of manufactured artefact, a variety of documents, many in the form of technical drawings, will be needed to convey to those responsible for constructing the design the information necessary to be able to proceed according to the wishes of the designer. Such sets of drawings may be passed between the people or companies responsible for the construction, enabling all those involved to make adjustments or suggest changes to the design. In some cases there may well be a considerable number of drawings required in such sets of drawings. In AutoCAD 2005 all the drawings from which a design is to be manufactured can be gathered together in a **sheet set**. This chapter shows how a much reduced sheet set of drawings for the construction of a house at 62 Pheasant Drive can be formed. Some other drawings, particularly detail drawings, would be required in this example, but to save page space, the sheet set described here consists of only four drawings and a subset of another four.

Example – a sheet set for 62 Pheasant Drive

1. Construct a template **62 Pheasant Drive.dwt** based upon the **acadiso.dwt** template, but including a border and a title block. Save the template in a **Layout1** format. An example of the title block from one of the drawings constructed in this template is shown in Fig. 11.1.
2. Construct each of the drawings which will form the sheet set in this drawing template. The whole set of drawings is shown in Fig. 11.10 on

Fig. 11.1 The title block from Drawing number **2** of the sheet set

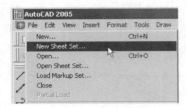

Fig. 11.2 Selecting **New Sheet Set** . . . from the **File** drop-down menu

page 230. Save the drawings in a directory – in this example this has been given the name **New building**.

3. *Click* **New Sheet Set** in the **File** drop-down menu (Fig. 11.2). The first of a series of **Create Sheet Set** dialogs appears. *Click* the radio button next to **Existing drawings**, followed by a *click* on the **Next** button and the next dialog appears (Fig. 11.3).

4. *Enter* details as shown in the dialog in Fig. 11.3. Then *click* the **Sheet Set Properties** button. The **Sheet Set Properties** dialog appears showing the properties of the proposed sheet set (Fig. 11.4).

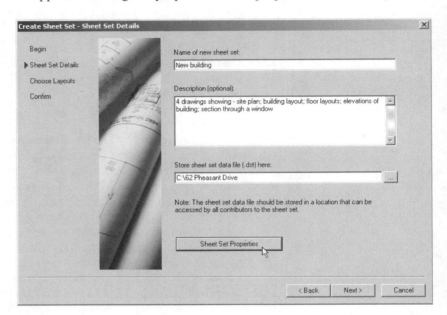

Fig. 11.3 The first of the **Create Sheet Set** dialogs

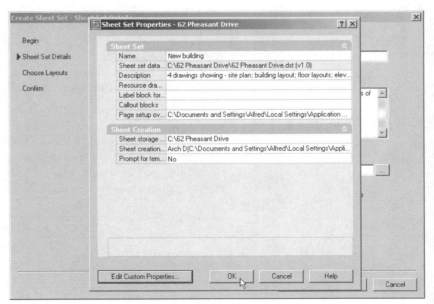

Fig. 11.4 The **Sheet Set Properties** dialog

If satisfied with the details in the dialog, *click* its **OK** button. The **Create Sheet Set** dialog reappears. *Click* its **Next** button to call the third dialog to screen (Fig. 11.3).

5. *Click* its **Browse** button and from the **Browse for Folder** list which comes to screen, *pick* the directory **New building**. *Click* the **OK** button and the drawings held in the directory appears in the **Choose Layouts** dialog (Fig. 11.5). If satisfied that the list is correct, *click* the **Next** button. A **Confirm** dialog appears. If satisfied *click* the **Finish** button and the **Sheet Set Manager** palette appears showing the drawings which will be in the **New building** sheet set (Fig. 11.6).

Fig. 11.5 The **Choose Layouts** dialog

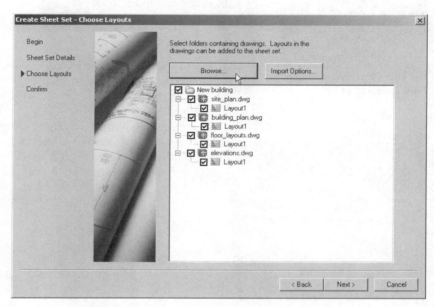

Fig. 11.6 The **Sheet Set Manager** palette for the **New building** sheet set

6. *Right-click* **New building** in the **Sheet Set Manager** and *left-click* **New Subset** . . . in the *right-click* menu which appears (Fig. 11.7). A dialog **Sub Set Properties** appears from which drawings for the sub set can be selected from the directory holding the sub set drawings.

Fig. 11.7 The *right-click* menu from the sheet name

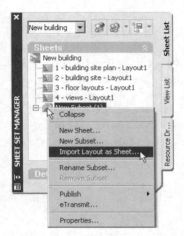

Fig. 11.8 The *right-click* menu from **New Sub Set (1)**

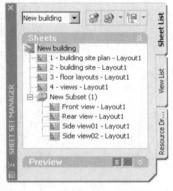

Fig. 11.9 The completed list of layout drawings in the sheet set

7. **Sub Set (1)** appears in the list of the **Sheet Set Manager**. *Right-click* **Sub Set (1)** and from the menu which appears select **Import Layout as Sheet** . . . (Fig. 11.8).

8. A dialog **Import Layout as Sheet** appears from which drawings can be selected one by one. Only drawings which have been saved as layouts (using the **Layout1** or **Layout2** tabs) can be added to a sub set. Add the four layout drawings as shown in Fig. 11.9.

9. *Double-click* on each of the icons to the left of the drawing names in the **Sheet Set Manager** and the drawings appear on screen. Fig. 11.10 shows the complete set and sub set of the eight drawings in the **New building** sheet set.

   Notes

1. If any of the eight drawings in the sheet set are subsequently amended or changed, when the drawings are opened again from the **New building** Sheet Set Manager the drawing will include any changes or amendments.

2. Drawings can be placed into sheet sets only if they have been saved in a **Layout** screen. Note that all the drawings shown in the **New building** Sheet Set Manager have **Layout1** after the drawing names because each has been saved after being placed in a **Layout1** screen.

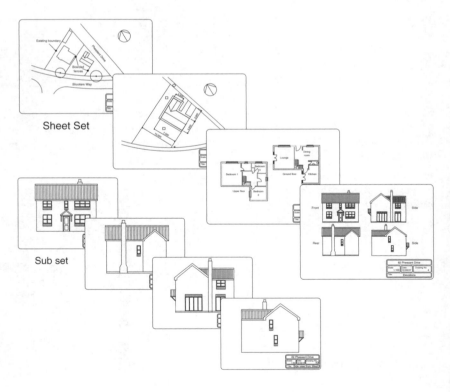

Fig. 11.10 The eight drawings in the **New building** sheet set

3. Sheet sets in the form of **DWF** (Design Web Format) files can be sent via email to others who are using the drawings or placed on an intranet. The method of producing a **DWF** for the **New building** sheet set follows.

New building DWF

1. In the **New building** Sheet Set Manager *click* the **Publish to Web** icon (Fig. 11.11). The **Select DWF File** dialog appears (Fig. 11.12). *Enter* **New building** in the **File name** field followed by a *click* on the **Select** button. An icon in the bottom right-hand corner of the AutoCAD 2005 window starts fluctuating in shape showing that the DWF file is being processed. When the icon becomes stationary again a warning balloon appears stating that the **Publish Job** is completed.

Fig. 11.11 The **Publish to Web** icon

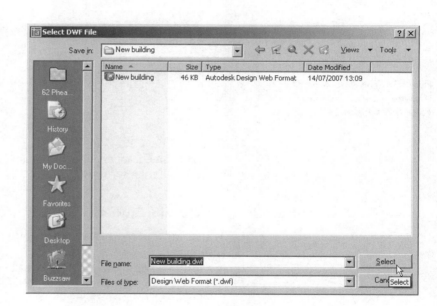

Fig. 11.12 The **Select DWF File** dialog

2. *Right-click* on the icon at the bottom right-hand corner of the AutoCAD 2005 window. The *right-click* menu (Fig. 11.13) appears.

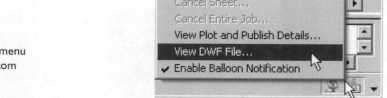

Fig. 11.13 The *right-click* menu from the icon at the bottom right-hand corner of the AutoCAD 2005 window

3. *Click* **View DWF File** . . . from the *right-click* menu. The **Autodesk DWF Viewer** window appears showing the **New building.dwf** file (Fig. 11.14). *Click* on any of the icons of the thumbnails of the drawings in the viewer and the drawing appears in the right-hand area of the viewer.

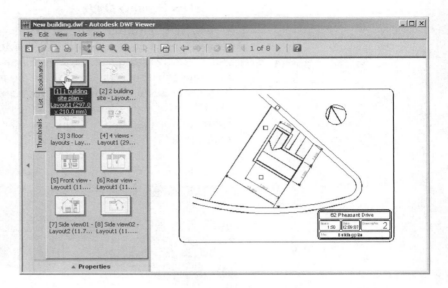

Fig. 11.14 The **Autodesk DWF Viewer** showing details of the **New building.dwf** file

4. If required the DWF Viewer file can be sent between people by email as an attachment, opened in a company's intranet or, indeed, included within an Internet web page.

Revision notes

1. To start off a new sheet set, *click* **New Sheet Set** . . . in the **File** drop-down menu.
2. Sheet sets can only contain drawings saved in **Layout** form.
3. Sheet sets can be published as **Design Web Format** (*.dwf) files which can be sent between offices by email, published on an intranet or published on a web page.
4. Sub sets can be included in sheet sets.
5. Changes or amendments made to any drawings in a sheet set are reflected in the sheet set drawings when the sheet set is opened.

Exercises

1. Fig. 11.15 is an exploded orthographic projection of the parts of a piston and its connecting rod. There are four parts in the assembly. Small drawings of the required sheet set are shown in Fig. 11.16.

   Construct the drawing in Fig. 11.15 and also the four drawings of its parts. Save each of the drawings in a **Layout1** format and construct the sheet set which contains the five drawings.

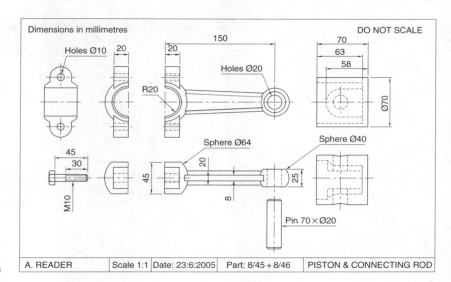

Fig. 11.15 Exercise 1 – the exploded orthographic projection

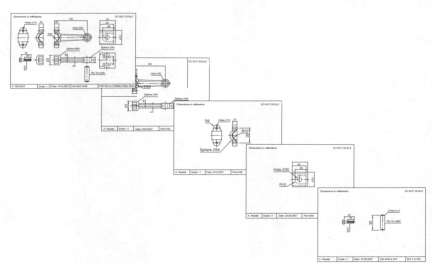

Fig. 11.16 Exercise 1 – the five drawings in the sheet set

Construct the **DWF** file of the sheet set. Experiment sending it to a friend via email as an attachment to a document, asking them to return the whole email to you without changes. When the email is returned, open its DWF file and *click* each drawing icon in turn to check the contents of the drawings.

Note

Fig. 11.17 shows a DWF for the sheet set from exercise 1 with the addition of a sixth drawing which is a 3D exploded model drawing of the five parts of the piston and connecting rod which have been Gourand shaded – see Chapter 16. This illustration has been included here to show that such shaded 3D models can be included in a sheet set.

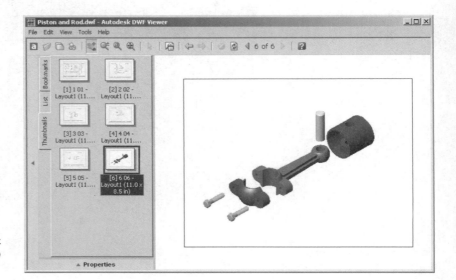

Fig. 11.17 The DWF of the sheet set which includes the shaded 3D model

2. Construct a similar sheet set as in the answer to exercise 1 from the exploded orthographic drawing of a **Machine adjusting spindle** given in Fig. 11.18.

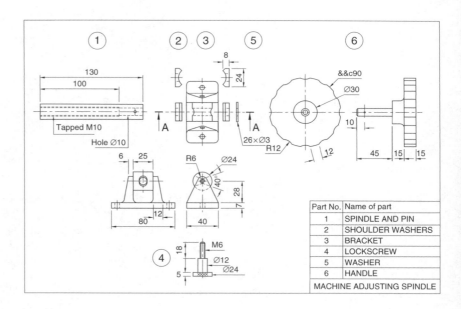

Fig. 11.18 Exercise 2

CHAPTER 12

Building drawing

Aim of this chapter

To show that AutoCAD 2005 is a suitable CAD software package for the construction of building drawing.

Building drawings

There are a number of different types of drawings related to the construction of any form of building. As fairly typical examples of a set of building drawings, in this chapter, seven drawings are shown related to the construction of an extension to an existing two-storey house (44 Ridgeway Road). These show:

1. A site plan of the original two-storey house, drawn to a scale of **1:200** (Fig. 12.1).

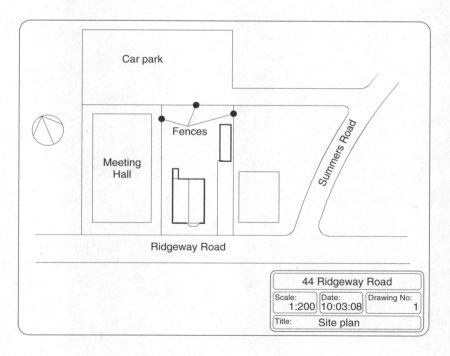

Fig. 12.1 A site plan

2. A site layout plan of the original house, drawn to a scale of **1:100** (Fig. 12.2).

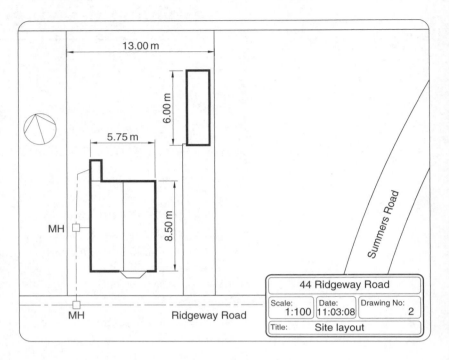

Fig. 12.2 A site layout plan

3. Floor layouts of the original house, drawn to a scale of **1:50** (Fig. 12.3).

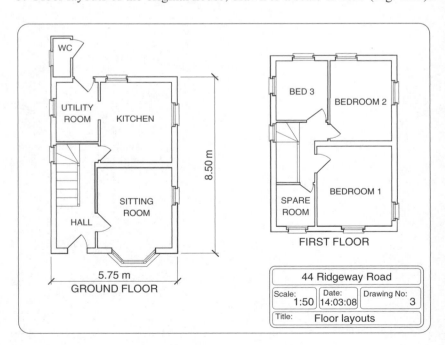

Fig. 12.3 Floor layouts of the original house

4. Views of all four sides of the original house, drawn to a scale of **1:50** (Fig. 12.4).

Fig. 12.4 Views of the original house

5. Floor layouts including the proposed extension, drawn to a scale of **1:50** (Fig. 12.5).

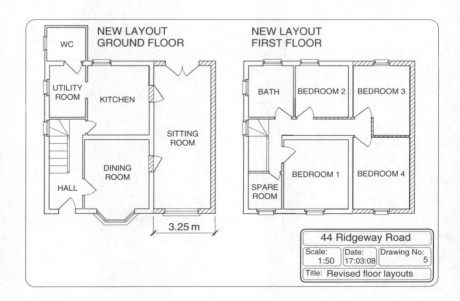

Fig. 12.5 Floor layouts of the proposed extension

6. Views of all four sides of the house including the proposed extension, drawn to a scale of **1:50** (Fig. 12.6).

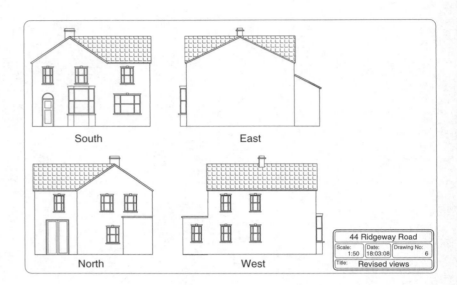

Fig. 12.6 Views including the proposed extension

7. A sectional view through the proposed extension, drawn to a scale of **1:50** (Fig. 12.7).

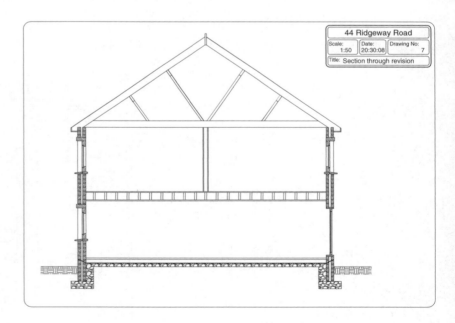

Fig. 12.7 A section through the proposed extension

Notes

1. Other types of drawings will be constructed such as drawings showing the details of parts such as doors, windows, floor structures, etc. These are often shown in sectional views.

2. Although the seven drawings related to the proposed extension of the house at 44 Ridgeway Road are shown here as having been constructed on either A3 or A4 layouts, it is common practice to include several types of building drawings on larger sheets such as A1 sheets of a size 820 mm by 594 mm.

Floor layouts

When constructing floor layout drawings it is advisable to build up a library of block drawings of symbols representing features such as doors, windows, etc. These can then be inserted into layouts from the DesignCenter. A suggested small library of such block symbols is shown in Fig. 12.8.

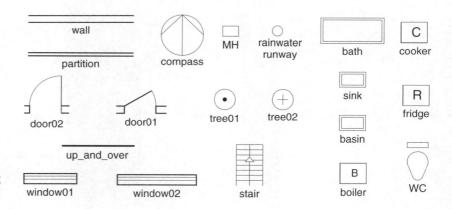

Fig. 12.8 A small library of building symbols

Revision notes

There are a number of different types of building drawings – site plans, site layout plans, floor layouts, views, sectional views, detail drawings, etc. AutoCAD 2005 is a suitable CAD program to use when constructing building drawings.

Exercises

1. Fig. 12.9 is a site plan drawn to a scale of 1:200 showing a bungalow to be built in the garden of an existing bungalow.

   Construct the library of symbols shown in Fig. 12.8 and by inserting the symbols from the DesignCenter construct a scale 1:50 drawing of the floor layout plan of the proposed bungalow.

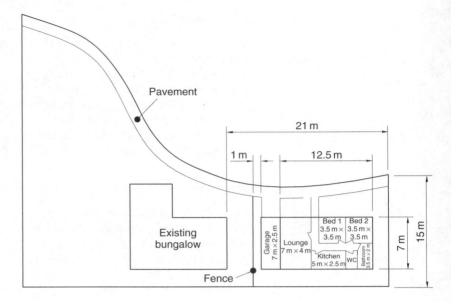

Fig. 12.9 Exercise 1

2. Fig. 12.10 is a site plan of a two-storey house to be built on a building plot.

Design and construct to a scale 1:50, a suggested pair of floor layouts for the two floors of the proposed house.

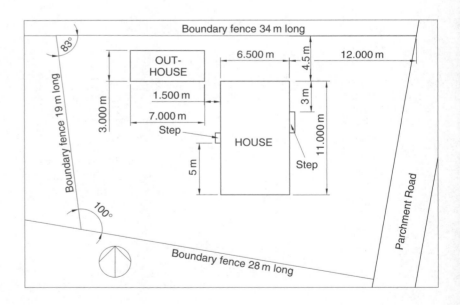

Fig. 12.10 Exercise 2

3. Fig. 12.11 shows a scale 1:100 site plan for the proposed bungalow at 4 Caretaker Road. Construct the floor layout for the proposed house as shown in the drawing in Fig. 12.12.

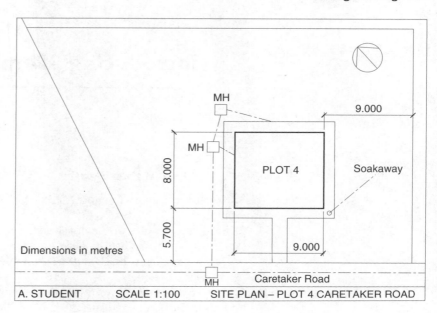

Fig. 12.11 Exercise 3 – site plan

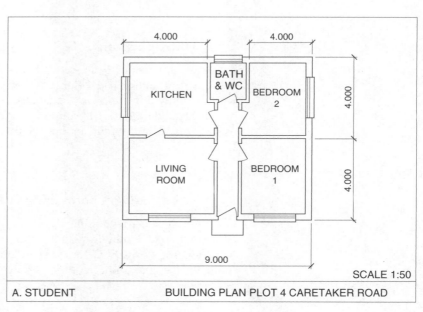

Fig. 12.12 Exercise 3

CHAPTER 13

Introducing 3D modelling

Aims of this chapter

1. To introduce all the tools used for the construction of 3D solid models.
2. To give examples of the construction of 3D solid models using all the tools from the **Solids** toolbar.
3. To give examples of 2D outlines suitable as a basis for the construction of 3D solid models.
4. To give examples of constructions involving the Boolean operators – **Union**, **Subtract** and **Intersect**.

Introduction

As shown in Chapter 1 the AutoCAD coordinate system includes a third coordinate direction **Z**, which has not been used so far when dealing with 2D drawings in previous chapters. 3D model drawings make use of this third **Z** coordinate. 3D solid model drawings are mainly constructed using tools from the four toolbars **Solids**, **Solids Editing**, **Surfaces** and **Render**.

The toolbars containing Solid and Render tools

The tools in the four toolbars containing the majority of the tools for constructing and rendering 3D solids and surfaces described in this book are shown in Fig. 13.1. When calling the tools from these four toolbars, the four same methods apply as are used when constructing 2D drawings:

1. A *click* on a tool icon in a toolbar brings the selected tool into action.
2. A *click* on the name of a tool from a drop-down menu brings the tool into action.
3. *Entering* the tool name at the command line in the command window, followed by pressing the *Return* button of the mouse or the **Return** key of the keyboard, brings the tool into action.
4. Some of the 3D tools have an abbreviation which can be *entered* at the command line instead of its full name.

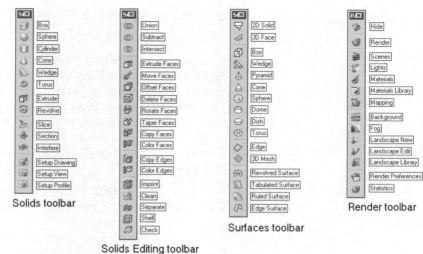

Solids toolbar

Solids Editing toolbar

Surfaces toolbar

Render toolbar

Fig. 13.1 The four toolbars with their tool icons

Notes

1. When constructing 2D drawings no matter which method is used, most operators will use a variety of these four methods. Calling a tool results in prompts sequences appearing at the command prompt as in the following example:

Command: *enter* **box** *right-click*
Specify corner of box or [CEnter]: *enter* **90,120** *right-click*
Specify corner or [Cube/Length]:

Or, if the tool is called from an icon or from a drop-down menu:

Command:_box
Specify corner of box or [CEnter]: *enter* **90,120** *right-click*
Specify corner or [Cube/Length]:

2. In the chapters which follow, once the prompts sequences for a tool are given as in the example above, if the tool's sequences are to be repeated, they may be replaced by an abbreviated form such as:

Command: box
[prompts]: 90,120
[prompts]:

Examples of 3D drawings using the 3D Face tool

The following two examples demonstrate the construction of 3D faces using the **3D Face** tool from the **Surfaces** toolbar. A 3D face is a triangular or quadrilateral flat (planar) or non-planar surface – i.e. a surface with either three or four edges in 3D space. Lines or other surfaces behind 3D faces can be hidden by using the **Hide** tool.

First example – 3D Face tool (Fig. 13.2)

1. At the command line:

   Command: *enter* **3dface** *right-click*
   Specify first point or [Invisible]: *enter* **60,230** *right-click*
   Specify second point or [Invisible]: *enter* **60,110** *right-click*
   Specify third point or [Invisible] <exit>: *enter* **190,110,150** *right-click*
   Specify fourth point or [Invisible] <create three-sided face>: *enter*
       190,230,150 *right-click*
   Specify third point or [Invisible] <exit>: *right-click*
   Command:

2. Call the **Mirror** tool and mirror the 3D face about **190,110** and **190,230**.
3. *Click* on **View** in the menu bar and on **3D Views** in the drop-down menu which appears, and *click* again on **SW Isometric** in the sub-menu which comes on screen alongside the drop-down menu.
4. *Click* on the **Hide** tool from the **View** drop-down menu.
   After **HIDE**, part of the right-hand 3D face has been hidden behind the left-hand face.

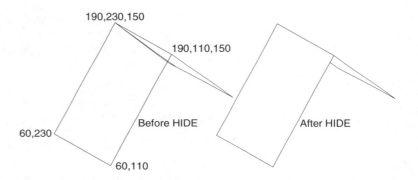

Fig. 13.2 First example – **3D Face tool**

Second example – 3D Face tool (Fig. 13.3)

It is assumed in this example that the reader understands how to *enter* 3D coordinates and how to place the screen in various **3D Views** from the **View** drop-down menu.

1. *Click* on **View** in the menu bar and again on **3D Views** in the drop-down menu. *Click again* on **Front** in the sub-menu which appears.
2. Call **Zoom** and zoom to **1**.
3. Call the **3D Face** tool:

   Command: 3dface
   [prompts]: 130,0
   [prompts]: 250,0
   [prompts]: i (for Invisible)
   [prompts]: 250,110
   [prompts]: 130,110
   [prompts]: 190,250

[**prompts**]: 250,110
[**prompts**]: *right-click*
Command:

4. *Click* on **View** in the menu bar and again on **3D Views** in the drop-down menu. *Click* again on **Right** in the sub-menu which appears.
5. Call **Zoom** and zoom to **1**.
6. With **Mirror**, mirror the face about the line **320,0** and **320,160**.
7. *Click* on **View** and on **3D Views** and select **SW Isometric** from the sub-menu. **Zoom** to **1**.
8. Call **3D Face** and with the aid of the osnap **endpoint** construct four 3D faces linking the two previously constructed faces.
9. Call **Hide** from the **View** drop-down menu.

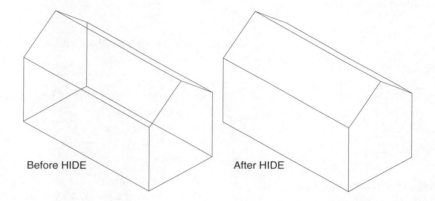

Before HIDE   After HIDE

Fig. 13.3 Second example – **3D Face tool**

2D outlines suitable for 3D models

When constructing outlines suitable as a base for the construction of some forms of 3D model, they can be drawn using either the **Line** tool or the **Polyline** tool. If constructed with the **Line** tool, before being of any use for 3D modelling, the outline must be changed to a region using the **Region** tool.

First example – Line outline and Region (Fig. 13.4)

1. Call the **Line** tool from the **Draw** toolbar. Construct the outline as shown in the left-hand drawing of Fig. 13.4.
2. Copy the outline to produce a second line outline (right-hand drawing of Fig. 13.4).
3. At the command line:

Command: *enter* **pe** (Edit Polyline) *right-click*
PEDIT Select polyline or [Multiple]: *enter* **m** (Multiple) *right-click*
Select objects: window the right-hand outline **12 found**
Select objects: *right-click*
Convert lines and arcs to polylines [Yes/No] <Y>: *right-click*

Enter an option [prompts]: *enter* **j** (Join) *right-click*
12 segments added to polyline.
Command:

And the **Line** outline is changed to a polyline.

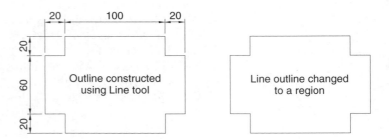

Fig. 13.4 First example – **Line outline and Region**

Fig. 13.5 The **Region** tool icon in the **Draw** toolbar

4. *Left-click* on the **Region** tool in the **Draw** toolbar (Fig. 13.5). The command line shows:

Command:_region
Select objects: window the left-hand drawing **12 found**
Select objects: *right-click*
1 loop extracted
1 Region created
Command:

And the **Line** outline is changed to a region. Either of the outlines can now be used when constructing some types of 3D model.

Second example – Union, Subtract and regions (Fig. 13.6)

1. Construct drawing **1** of Fig. 13.6 and with the **Copy Object** tool, copy the drawing three times to produce drawings **2**, **3** and **4**.
2. With the **Region** tool change all the outlines into regions.

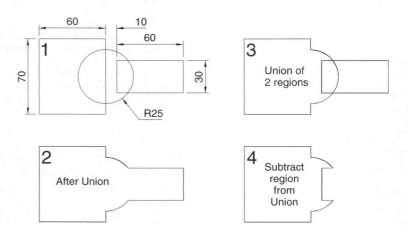

Fig. 13.6 Second example – **Union, Subtract and regions**

Fig. 13.7 The **Union** tool icon from the **Solids Editing** toolbar

3. Drawing **2** – call the **Union** tool from the **Solids Editing** toolbar (Fig. 13.7). The command line shows:

Command:_union
Select objects: *pick* the left-hand region **1 found**
Select objects: *pick* the circular region **1 found, 2 total**
Select objects: *pick* the right-hand region **1 found, 3 total**
Command:

4. Drawing **3** – with the **Union** tool form a union of the left-hand region and the circular region.
5. Drawing **4** – call the **Subtract** tool from the **Solids Editing** toolbar (Fig. 13.8). The command line shows:

Fig. 13.8 The **Subtract** tool icon from the **Solids Editing** toolbar

Command:_subtract Select solids and regions to subtract from . . .
Select objects: *pick* the region just formed **1 found**
Select objects: *right-click*
Select solids and regions to subtract: *pick* the right-hand region **1 found**
Select objects: *right-click*
Command:

Third example – Intersection and regions (Fig. 13.9)

1. Construct drawing **1** of Fig. 13.9.
2. With the **Region** tool, change the three outlines into regions.

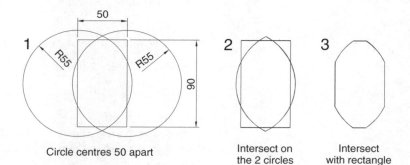

Fig. 13.9 Third example – **Intersection and regions**

Circle centres 50 apart

Intersect on the 2 circles

Intersect with rectangle

3. With the **Copy Object** tool, copy the three regions.
4. Drawing **2** – call the **Intersect** tool from the **Solids Editing** toolbar (Fig. 13.10). The command line shows:

Command:_intersect
Select objects: *pick* one of the circles **1 found**
Select objects: *pick* the other circle **1 found, 2 total**
Select objects: *right-click*
Command:

Fig. 13.10 The **Intersect** tool icon from the **Solids Editing** toolbar

And the two circular regions intersect with each other to form a region.
5. Drawing **3** – repeat using the **Intersect** tool on the intersection of the two circles and the rectangular region.

The Extrude tool

The **Extrude** tool can be called either, with a *click* on its tool icon in the **Solids** toolbar (Fig. 13.11), or by *entering* **extrude** or its abbreviation **ext** at the command line.

Fig. 13.11 The **Extrude** tool icon from the **Solids** toolbar

Fig. 13.12 Calling **Extrude** from the **Draw** drop-down menu

Extrude can also be called from the **Draw** toolbar – *click* on **Draw** in the menu bar and when the **Draw** drop-down menu appears, *click* on **Solids** in the drop-down menu and again on **Extrude** in the sub-menu which then appears (Fig. 13.12).

Examples of the use of the Extrude tool

The three examples of forming regions given in Figs 13.4, 13.6 and 13.9 are used here to show the results of using the **Extrude** tool.

First example – Extrude (Fig. 13.14)

From the first example of forming a region:

1. Call the **Extrude** tool. The command line shows:

   Command:_extrude
   Current wire frame density: ISOLINES=4
   Select objects: *pick* the region **1 found**
   Select objects: *right-click*
   Specify height of extrusion or [Path]: *enter* 50 *right-click*
   Specify angle of taper for extrusion <0>: *right-click*
   Command:

2. *Click* **View** in the menu bar, followed by a *click* on **3D Views** in the drop-down menu which appears, followed by another *click* on **SW Isometric** in the **3D Views** sub-menu (Fig. 13.13). The extrusion appears in an isometric view.
3. Call **Zoom** and zoom to **1**.
4. At the command line:

   Command: *enter* **hi** (abbreviation for Hide) *right-click*
   HIDE Regenerating model
   Command:

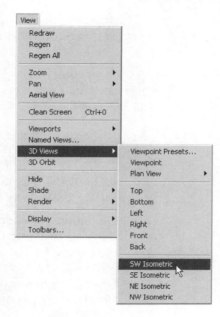

Fig. 13.13 The **3D Views** sub-menu from the **View** drop-down menu

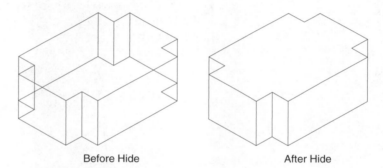

Fig. 13.14 First example – **Extrude**

Before Hide   After Hide

Notes

1. In the above example we made use of one of the isometric views possible using the **3D Views** sub-menu of the **View** drop-down menu. These views from this sub-menu will be used frequently in examples to show 3D solid model drawings in a variety of positions in 3D space.
2. Note also the use of the **Hide** tool. Extruded polylines or regions are made up from 3D faces (3D meshes). When **Hide** is called, lines behind the 3D meshes become invisible on screen.
3. **Hide** can also be called from the **View** drop-down menu, but the quickest method of calling the tool is to *enter* **hi** at the command line, followed by a *right-click*.
4. Note the **Current wire frame density: ISOLINES=4** in the prompts sequence when **Extrude** is called. The setting of **4** is suitable when

extruding plines or regions consisting of straight lines, but when arcs are being extruded it may be better to set **ISOLINES** to a higher figure as follows:

Command: *enter* isolines *right-click*
Enter new value for ISOLINES <4>: *enter* **16** *right-click*
Command:

Second example – Extrude (Fig. 13.15)

From the second example of forming a region:

1. Set **ISOLINES** to **16**.
2. Call the **Extrude** tool. The command line shows:

Command:_extrude
Current wire frame density: ISOLINES=16
Select objects: *pick* the region **1 found**
Select objects: *right-click*
Specify height of extrusion or [Path]: *enter* **50** *right-click*
Specify angle of taper for extrusion <0>: *enter* **5** *right-click*
Command:

3. *Click* **SW Isometric** in the **3D Views** sub-menu of the **View** drop-down menu (see Fig. 13.13).
4. **Zoom** to **1**.
5. Call **Hide**.

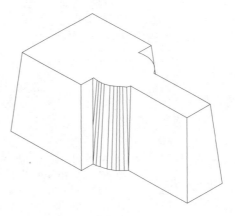

Fig. 13.15 Second example –
Extrude

Third example – Extrude (Fig. 13.16)

From the third example of forming a region:

1. Place the screen in the **3D Views/Front** view (see Fig. 13.13).
2. Construct a semicircular arc from the centre of the region.
3. Place the screen in the **3D Views/Top** view.
4. With the **Move** tool, move the arc to the centre of the region.
5. Place the screen in the **3D Views/SW Isometric** view.

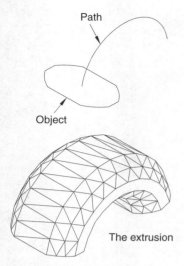

Path

Object

The extrusion

Fig. 13.16 Third example –
Extrude

Fig. 13.17 The **Revolve** tool icon
from the **Solids** toolbar

Fig. 13.18 First example –
Revolve. The closed pline

6. Set **ISOLINES** to **24**.

7. Call the **Extrude** tool. The command line shows:

Command:_extrude
Current wire frame density: ISOLINES=24
Select objects: *pick* the region **1 found**
Select objects: *right-click*
Specify height of extrusion or [Path]: *enter* **p** *right-click*
Select extrusion path or [Taper angle]: *pick* the path
Command:

The Revolve tool

The **Revolve** tool can be called either with a *click* on its tool icon in the **Solids** toolbar (Fig. 13.17), or by a *click* on its name in the **Solids** sub-menu of the **Draw** drop-down menu, or by *entering* **revolve**, or its abbreviation **rev**, at the command line.

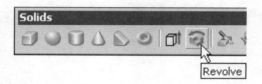

Examples of the use of the Revolve tool

Solids of revolution can be constructed from closed plines or from regions.

First example – Revolve (13.19)

1. Construct the closed polyline in Fig. 13.18.

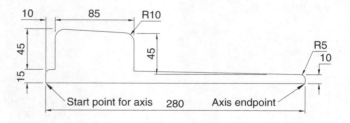

2. Set **ISOLINES** to **24**.

3. Call the **Revolve** tool. The command line shows:

Command:_revolve
Current wire frame density=24
Select objects: *pick* the polyline
Select objects: *right-click*
Specify start point for axis of revolution or define axis by
    [Object/X (axis)/Y (axis)]: *pick*

Specify endpoint of axis: *pick*
Specify angle of revolution <360>: *right-click*
Command:

4. Place in the **3D Views/SW Isometric** view.
5. Call **Hide**.

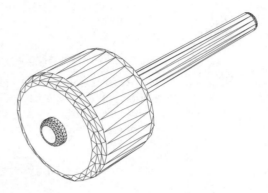

Fig. 13.19 First example –
Revolve

Second example – Revolve (Fig. 13.21)

1. Place the screen in the **3D Views/Front** view. **Zoom** to **1**.
2. Construct the pline outline (Fig. 13.20).
3 Set **ISOLINES** to **24**.
4. Call the **Revolve** tool and construct a solid of revolution.
5. Place the screen in the **3D Views/SW Isometric** view.
6. Call **Hide**.

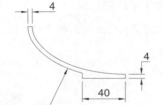

Semi-ellipse based
on 180×100 axes

Fig. 13.20 Second example –
Revolve. The pline outline

Fig. 13.21 Second example –
Revolve

Third example – Revolve (Fig. 13.23)

1. Construct the pline in Fig. 13.22. The drawing must be either a closed
   pline or a region.
2. Call **Revolve** and form a solid of revolution through 180°.

Fig. 13.22 Third example –
Revolve. The outline to be
revolved

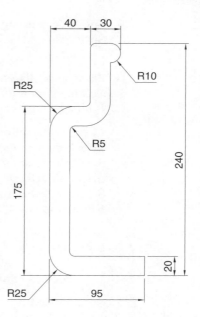

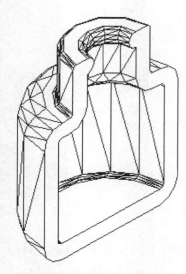

Fig. 13.23 Third example –
Revolve

Fig. 13.24 The **Box** tool icon
from the **Solids** toolbar

3D objects

At the command line:

> **Command:** *enter* **3d** *right-click*
> **Enter an option [Box/Cone/DIsh/DOme/Pyramid/Sphere/Torus/
> Wedge]:**

And the 3D objects names appear. Any one of the 3D objects can then be
called by *entering* the capital letter of the name of the 3D objects. The 3D
objects can also be called from the **Solids** toolbar or by *entering* the name
(e.g. **box**) at the command line.

First example – 3D Objects (Fig. 13.26)

1. Place the screen in the **3D Views/Front** view.
2. *Click* the **Box** tool icon in the **Solids** toolbar (Fig. 13.24). The command line shows:

> **Command:_box**
> **Specify corner of box or [Center] <0,0,0>:** *enter* **90,90** *right-click*
> **Specify corner or [Cube/Length]:** *enter* **110,−30** *right-click*
> **Specify height:** *enter* **75** *right-click*
> **Command:** *right-click*
> **BOX Specify corner of box or [Center] <0,0,0>:** 110,90
> **Specify corner or [Cube/Length]:** 170,70
> **Specify height:** 75
> **Command:** *right-click*
> **BOX Specify corner of box or [Center] <0,0,0>:** 110,−10

Fig. 13.25 The **Union** tool icon from the **Solids Editing** toolbar

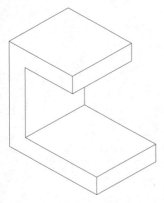

Fig. 13.26 First example – **3D Objects**

Fig. 13.27 The **Sphere** tool icon from the **Solids** toolbar

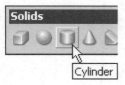

Fig. 13.28 The **Cylinder** tool icon from the **Solids** toolbar

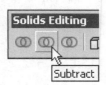

Fig. 13.29 The **Subtract** tool icon from the **Solids Editing** toolbar

Specify corner or [Cube/Length]: 200,−30
Specify height: 75
Command:

3. Place in the **3D Views/SE Isometric** view and **Zoom** to 1.
4. Call the **Union** tool from the **Solids Editing** toolbar (Fig. 13.25). The command line shows:

Command:_union
Select objects: *pick* one of the boxes **1 found**
Select objects: *pick* the second of box **1 found, 2 total**
Select objects: *pick* the third box **1 found, 3 total**
Select objects: *right-click*
Command:

And the three boxes are joined in a single union.
5. Call **Hide**.

Second example – 3D Objects (Fig. 13.30)

1. Set **ISOLINES** to **16**.
2. *Click* the **Sphere** tool icon from the **Solids** toolbar (Fig. 13.27). The command line shows:

Command:_sphere
Current wire frame density: ISOLINES=16
Specify center of sphere <0,0,0>: 180,170
Specify radius of sphere or [Diameter]: 50
Command:

3. *Click* the **Cylinder** tool icon in the **Solids** toolbar (Fig. 13.28). The command line shows:

Command:_cylinder
Current wire frame density: ISOLINES=16
Specify centre for base of cylinder or [Elliptical] <0,0,0>: 180,170
Specify radius for base of cylinder or [Diameter]: 25
Specify height of cylinder or [Center of other end]: 110
Command:

4. Place the screen in the **3D Views/Front** view and **Zoom** to **1**.
5. With the **Move** tool, move the cylinder vertically down so that the bottom of the cylinder is at the bottom of the sphere.
6. *Click* the **Subtract** tool icon in the **Solids Editing** toolbar (Fig. 13.29). The command line shows:

Command:_subtract Select solids and regions to subtract from
Select objects: *pick* the sphere **1 found**
Select objects: *right-click*
Select objects and regions to subtract: *pick* the cylinder
Select objects: *right-click*
Command:

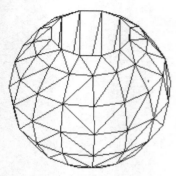

Fig. 13.30 Second example – **3D Objects**

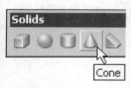

Fig. 13.31 The **Cone** tool icon from the **Solids** toolbar

Fig. 13.32 Third example – **3D Objects**

7. Place the screen in the **3D Views/SW Isometric** view and **Zoom** to **1**.
8. Call **Hide**.

Third example – 3D Objects (Fig. 13.32)

1. Call the **Cylinder** tool and with a centre **170,150** construct a cylinder of radius **60** and height **15**.
2. *Click* the **Cone** tool icon in the **Solids** toolbar (Fig. 13.31). The command line shows:

   Command:_cone
   Current wire frame density: ISOLINES=16
   Specify center point for base of cone or [Elliptical]: 170,150
   Specify radius for base of cone or [Diameter]: 40
   Specify height of cone or [Apex]: 150
   Command:

3. Call the **Sphere** tool and construct a sphere of centre **170,150** and radius **45**.
4. Place the screen in the **3D Views/Front** view and with the **Move** tool, move the cone and sphere so that the cone is resting on the cylinder and the centre of the sphere is at the apex of the cone.
5. Place in the **3D Views/SW Isometric** view and **Zoom** to **1** and with **Union** form a single 3D model from the three objects.
6. Call **Hide**.

Fourth example – 3D Objects (Fig. 13.33)

1. *Click* the **Box** tool icon in the **Solids** toolbar and construct two boxes – the first of corners **70,210** and **290,120** and of height **10**, the second of corners **120,200,10** and **240,130,10** and of height **80**.
2. Place the screen in the **3D Views/Front** view and **Zoom** to **1**.
3. *Click* the **Wedge** tool icon in the **Solids** toolbar. The command line shows:

   Command:_wedge
   Specify first corner of wedge or [Center]: 120,170,10
   Specify corner or [Cube/Length]: 80,160,10
   Specify height: 70
   Command: *right-click*
   WEDGE
   Specify first corner of wedge or [Center]: 240,170,10
   Specify corner or [Cube/Length]: 280,160,10
   Specify height: 70
   Command:

4. Place the screen in the **3D Views/SW Isometric** view and **Zoom** to **10**.
5. Call the **Union** tool from the **Solids Editing** toolbar and in response to the prompts in the tool's sequences *pick* each of the five objects in turn to form a union of the five objects.
6. Call **Hide**.

Fig. 13.33 Fourth example – **3D Objects**

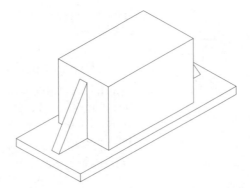

Fig. 13.34 The **Torus** tool icon from the **Solids** toolbar

Fig. 13.35 Fifth example – **3D Objects**

Fig. 13.36 The **Chamfer** and **Fillet** tool icons from the **Modify** toolbar

Fifth example – 3D Objects (Fig. 13.35)

1. Using the **Cylinder** tool from the **Solids** toolbar, construct a cylinder of centre **180,160**, of radius **40** and height **120**.
2. *Click* the **Torus** tool icon in the **Solids** toolbar (Fig. 13.34). The command line shows:

   Command:_torus
   Current wire frame density: ISOLINES=16
   Specify center point of torus <0,0,0>: 180,160,10
   Specify radius of torus or [Diameter]: 40
   Specify radius of tube or [Diameter]: 10
   Command: *right-click*
   TORUS
   Current wire frame density: ISOLINES=16
   Specify center point of torus <0,0,0>: 180,160,110
   Specify radius of torus or [Diameter]: 40
   Specify radius of tube or [Diameter]: 10
   Command:

3. Call the **Cylinder** tool and construct another cylinder of centre **180,160**, of radius **35** and height **120**.
4. Place in the **3D Views/SW Isometric** view and **Zoom** to **1**.
5. *Click* the **Union** tool icon in the **Solids Editing** toolbar and form a union of the larger cylinder and the two tori.
6. *Click* the **Subtract** tool icon in the **Solids Editing** toolbar and subtract the smaller cylinder from the union.
7. Call **Hide**.

The Chamfer and Fillet tools

The **Chamfer** and **Fillet** tools from the **Modify** toolbar (Fig. 13.36), which are used to create chamfers and fillets in 2D drawings in AutoCAD 2005, can just as well be used when constructing 3D models.

Example – Chamfer and Fillet (Fig. 13.39)

1. Using the **Box** and **Cylinder** tools, construct the 3D model in Fig. 13.37.
2. Place in the **3D Views/SW Isometric** view (Fig. 13.38). **Union** the two boxes and with the **Subtract** tool, subtract the cylinders from the union.

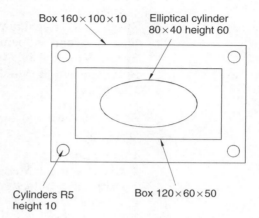

Box 160×100×10   Elliptical cylinder 80×40 height 60

Cylinders R5 height 10   Box 120×60×50

Fig. 13.37 Example – **Chamfer and Fillet** – the model before using the tools

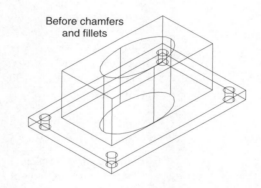

Before chamfers and fillets

Fig. 13.38 Example – isometric view – **Chamfer and Fillet** – the model before using the tools

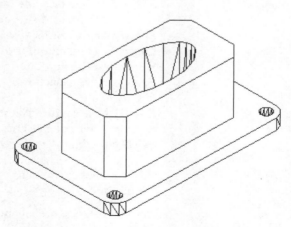

Fig. 13.39 Example – **Chamfer and Fillet**

Note

To construct the elliptical cylinder:

Command:_cylinder
Current wire frame density: ISOLINES=16
Specify centre for base of cylinder or [Elliptical] <0,0,0>: *enter* **e**
    right-click
Specify axis endpoint of ellipse for base of cylinder: 130,160
Specify second endpoint of ellipse for base of cylinder: 210,160
Specify length of axis for base of cylinder: 170,180
Specify height of cylinder or [Center of other end]: 50
Command:

3. *Click* the **Fillet** tool icon in the **Modify** toolbar (Fig. 13.36). The command line shows:

Command:_fillet
Current settings: Mode = TRIM. Radius = 1
Specify first object or [Polyline/Radius/Trim/mUltiple]: *enter* r
    (Radius) *right-click*
Specify fillet radius <1>: 10
Select first object: *pick* one corner
Select an edge: *pick* a second corner
Select an edge: *pick* a third corner
Select an edge: *pick* the fourth corner
Select an edge: *right-click*
4 edges selected for fillet.
Command:

4. *Click* the **Chamfer** tool icon in the **Modify** toolbar (Fig. 13.36). The command line shows:

Command:_chamfer
(TRIM mode) Current chamfer Dist1 = 1, Dist2 = 1
Select first line or [Polyline/Distance/Angle/Trim/Method/mUltiple]:
    enter **d** (Distance) *right-click*
Specify first chamfer distance <1>: 10
Specify second chamfer distance <1>: 10
Select first line: *pick* one corner – one side of the box highlights
Base surface selection . . . Enter surface selection <OK>: *right-click*
Specify base surface chamfer distance <10>: *right-click*
Specify other surface chamfer distance <10>: *right-click*
Select an edge: *pick* the edge again **Select an edge**: *pick* the second edge
Select an edge: *right-click*
Command:

And two edges are chamfered. Repeat to chamfer the other two edges.

Note on the tools Union, Subtract and Intersect

The tools **Union**, **Subtract** and **Intersect** found in the **Solids Editing** toolbar are known as the **Boolean** operators after the mathematician Boolean. They can be used to form unions, subtractions or intersections between extrusions, solids of revolution or any of the 3D Objects.

Note on using Modify tools on 3D models

As was seen above when using the **Move**, **Chamfer** and **Fillet** tools from the **Modify** toolbar, so also can other tools from the toolbar be used in connection with the construction of 3D models. The tools **Copy Object**, **Mirror**, **Rotate** and **Scale** from **Modify** can also be used when constructing 3D models if wished.

Note on rendering

Rendering of 3D models to produce photo-like images will be more fully described in pages 219 to 230. But to give more realism to some of the 3D models so far shown in this chapter, try the following for some of the models which have so far been constructed.

1. Open the file of one of the 3D models so far constructed.
2. At the command line:

   Command: *enter* **render** *right-click*

   And the **Render** dialog appears (Fig. 13.40). *Click* its **Render** button.

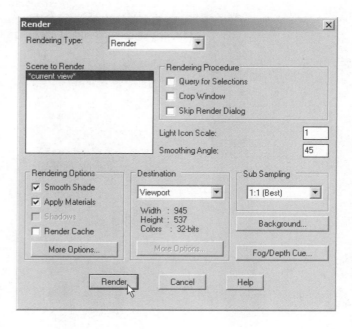

Fig. 13.40 The **Render** dialog

The model renders to a default set of rendering settings.

Examples of renderings of some of the models so far constructed are shown in Fig. 13.41.

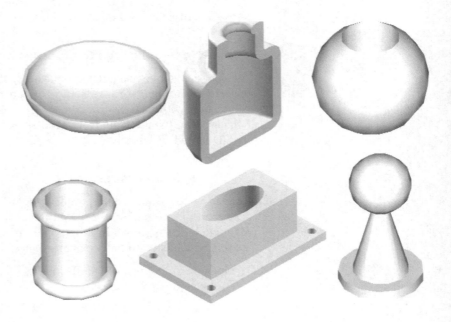

Fig. 13.41 Examples of simple renderings of 3D models

Note

3D models can be rendered because they are made up from a number of 3D meshes. Regions can also be rendered as shown in the renderings of some regions constructed earlier in this chapter (Fig. 13.42).

Fig. 13.42 Examples of renderings of regions

Revision notes

1. In the AutoCAD 3D coordinate system, positive Z is towards the operator away from the monitor screen.
2. A 3D face is a mesh behind which other details can be hidden.
3. The **Hide** tool can be used to hide details behind the 3D meshes of 3D solid models.
4. The **Extrude** tool can be used for extruding closed plines or regions to stated heights, to stated slopes or along paths.
5. The **Revolve** tool can be used for constructing solids of revolution through any angle up to 360°.
6. 3D models can be constructed from the 3D objects **Box**, **Sphere**, **Cylinder**, **Cone**, **Torus** and **Wedge**. Extrusions and/or solids of revolution may form part of models constructed using the 3D objects.
7. Tools such as **Chamfer** and **Fillet** from the **Modify** toolbar can be used when constructing 3D models.
8. A simple form of rendering of 3D models can be achieved by *entering* **render** at the command line followed by a *click* on the **Render** button of the dialog which appears.
9. The tools **Union**, **Subtract** and **Intersect** are known as the **Boolean** operators.

Exercises

The first three exercises given below give practice in the use of the **3D Face** and **Hide** tools. A variety of 3D viewing positions selected from **View/3D Views** will need to be used when constructing the 3D model drawings in answer to these three exercises.

1. Construct the four steps shown in Fig. 13.43 working to the given sizes.

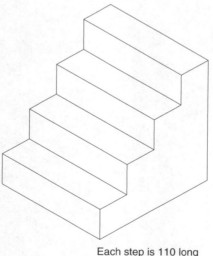

Each step is 110 long
by 30 deep and 30 high

Fig. 13.43 Exercise 1

2. Construct the 3D model given in Fig. 13.44 working to the sizes given with the drawing.

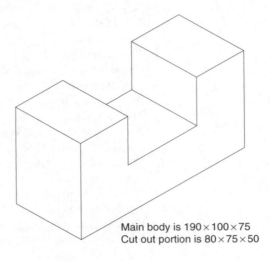

Main body is 190×100×75
Cut out portion is 80×75×50

Fig. 13.44 Exercise 2

3. Construct the 3D model drawing of a V block shown in Fig. 13.45 working to the sizes given in the right-hand drawing.

The exercises which follow require using the 3D objects tools and the tools **Revolve**, **Extrude**, **Union** and **Subtract**.

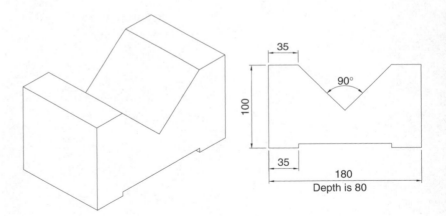

Fig. 13.45 Exercise 3

4. Fig. 13.46 shows a rendering of the solid of revolution for this exercise. Fig. 13.47 shows the outline from which the solid was obtained. Using the **Revolve** tool construct the solid of revolution and render the result.

Fig. 13.46 Exercise 4 –
a rendering

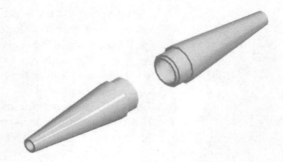

Fig. 13.47 Exercise 4

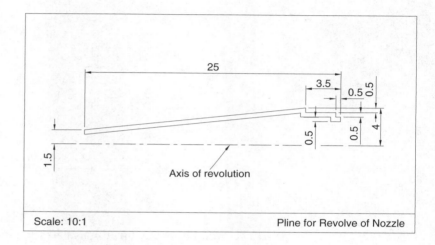

25

3.5

0.5

0.5

0.5

0.5

4

1.5

Axis of revolution

Scale: 10:1   Pline for Revolve of Nozzle

5. Construct the rendered 3D model of a wine glass as shown in Fig. 13.48, working to the dimensions given in the outline drawing (Fig. 13.49).

Fig. 13.48 Exercise 5 –
a rendering

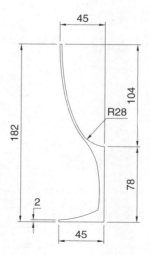

45

182

104

R28

78

2

45

Fig. 13.49 Exercise 5

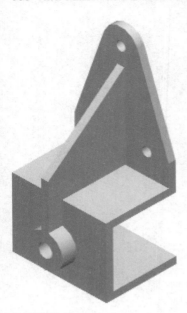

Fig. 13.50 Exercise 6 – A rendering

Fig. 13.51 Exercise 6

You will need to construct the outline and change it into a region before being able to change the outline into a solid of revolution using the **Revolve** tool. This is because the semi-elliptical part of the outline has been constructed using the **Ellipse** tool, resulting in part of the outline being a spline, which cannot be acted upon by **Polyline Edit** to form a closed pline.

6. Fig. 13.50 is a rendering of the 3D solid model constructed by working to the information given in Fig. 13.51. Construct the bracket and render as shown in Fig. 13.50.

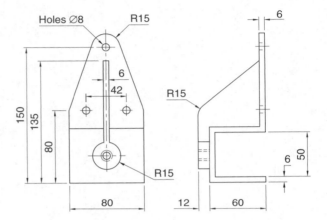

7. Working to the dimensions given in Fig. 13.52, construct an extrusion of the plate to a height of **5** units and render as shown, in Fig. 13.53.

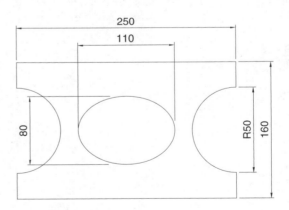

Fig. 13.52 Exercise 7

8. Working to the details given in the orthographic projection in Fig. 13.54, construct a 3D model of the assembly as shown in Fig. 13.55.

After constructing the pline outline(s) required for the solid(s) of revolution, use the **Revolve** tool to form the 3D solid.

Fig. 13.53 Exercise 7 – a
rendering

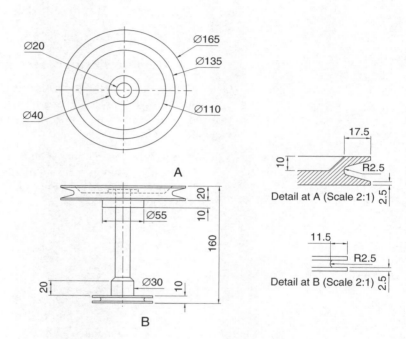

A

B

Detail at A (Scale 2:1)

Detail at B (Scale 2:1)

Fig. 13.54 Exercise 8

Fig. 13.55 Exercise 8 –
a rendering

CHAPTER 14

3D models in viewports

Aim of this chapter

To give examples of 3D solid models constructed in multiple viewport settings.

Setting up viewport systems

One of the better methods of constructing 3D models is in different viewport settings. Working in viewport settings allows what is being constructed to be seen from a variety of viewing positions. To set up a new viewport system:

1. *Click* **View** in the menu bar and from the drop-down menu which appears *click* **Viewports** and in the sub-menu which then appears *click* **New Viewports** . . . (Fig. 14.1). The **Viewports** dialog appears (Fig. 14.2).

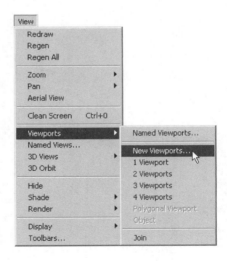

Fig. 14.1 Selecting **New Viewports** . . . from the **View** drop-down menu

2. *Click* the **New Viewports** tab and a number of named viewports systems appear in the **Standard viewports** list in the dialog.
3. *Click* the name **Four: Equal**, followed by a *click* on **3D** in the **Setup** popup list. A preview of the **Four: Equal** viewports screen appears showing the views appearing in each of the four viewports.

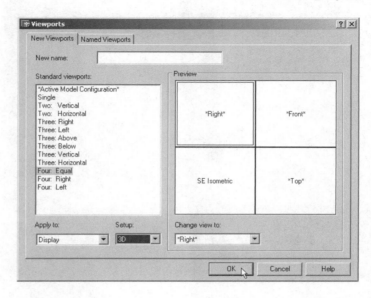

Fig. 14.2 The **Viewports** dialog

4. *Click* the **OK** button of the dialog and the AutoCAD 2005 drawing area appears showing the four viewport layout (Fig. 14.3).

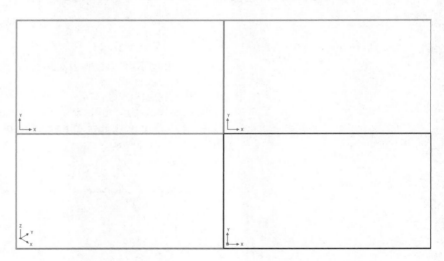

Fig. 14.3 The **Four: Equal** viewports layout

First example – Four: Equal viewports (Fig. 14.7)

Fig. 14.4 shows a first angle orthographic projection of a support. To construct a **Scale 1:1** 3D model of the support in a **Four: Equal** viewport setting:

1. *Click* **View** in the menu bar, followed by a *click* on **Viewports** in the drop-down menu, followed by another *click* on **New Viewports** . . . in the **Viewports** sub-menu. Make sure the **3D** option is selected from the **Setup** popup list and *click* the **OK** button of the dialog. The AutoCAD 2005 drawing area appears in a **Four: Equal** viewport setting.

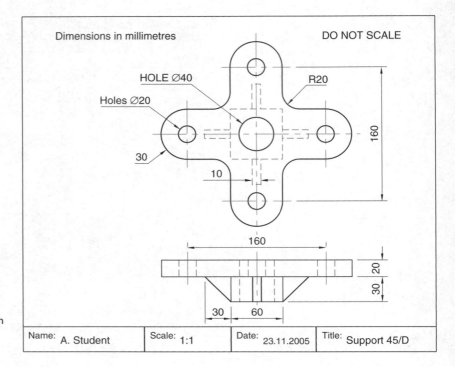

Fig. 14.4 Orthographic projection of the support for the first example

2. *Click* in the **Top** viewport (bottom right-hand corner viewport) to make it current.
3. Set **ISOLINES** to **4**.
4. Using the **Polyline** tool, construct the outline of the plan view of the plate of the support, including the holes (Fig. 14.5). Note the views in the other viewports.

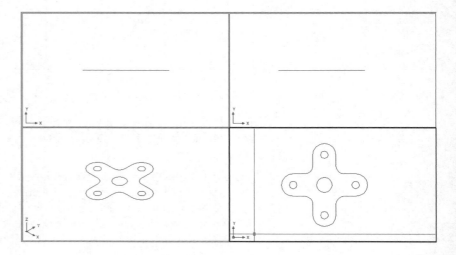

Fig. 14.5 The plan view drawn

5. Call the **Extrude** tool from the **Solids** toolbar and extrude the plan outline and the circles to a height of **20**.
6. With the **Subtract** tool from the **Solids Editing** toolbar, subtract the holes from the plate (Fig. 14.6).

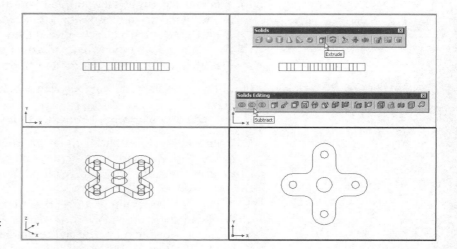

Fig. 14.6 The four views after using the **Extrude** and **Subtract** tools

7. Call the **Box** tool and in the centre of the plate construct a box of Width = **60**, Length = **60** and Height = **30**.
8. Call the **Cylinder** tool and in the centre of the box construct a cylinder of Radius = **20** and Height = **30**.
9. Call **Subtract** and subtract the cylinder from the box.
10. *Click* in the **Right** viewport. With the **Move** tool, move the box and its hole into the correct position with regard to the plate.

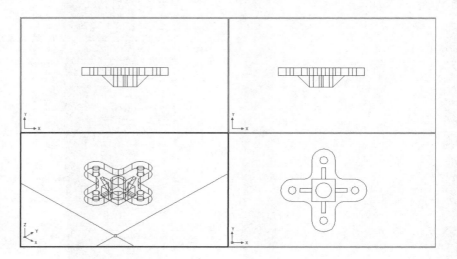

Fig. 14.7 First example – **Four: Equal viewports**

11. With **Union**, form a union of the plate and box.

12. *Click* in the **Front** viewport and construct a triangle of one of the webs attached between the plate and the box. With **Extrude**, extrude the triangle to a height of **10**. With the **Mirror** tool, mirror the web to the other side of the box.

13. *Click* in the **Right** viewport and with the **Move** tool, move the two webs into their correct position between the box and the plate. Then, with **Union**, form a union between the webs and the 3D model.

14. While in the **Right** viewport, construct the other two webs and in the **Front** viewport, move, mirror and union the webs as in step **13**.

Fig. 14.6 shows the resulting four-viewport scene.

Second example – Four: Left viewports (Fig. 14.9)

1. Open the **Four: Left** viewport layout from the **Viewports** dialog.

2. In the **Top** viewport construct an outline of the web of the Support Bracket shown in Fig. 14.8. With the **Extrude** tool, extrude the parts of the web to a height of **20**.

3. With the **Subtract** tool, subtract the holes from the web.

4. While in the **Top** viewport, construct two cylinders central to the extrusion – one of radius **50** and height **30**, the second of radius **40** and height **30**. With the **Subtract** tool, subtract the smaller cylinder from the larger.

5. *Click* in the **Front** viewport and move the cylinders vertically upwards by **5** units. Then with **Union** form a union between the cylinders and the web.

6. Make the **Front** viewport active and at one end of the union, construct two cylinders – the first of radius **10** and height **80**, the second of radius **15** and height **80**. Subtract the smaller from the larger.

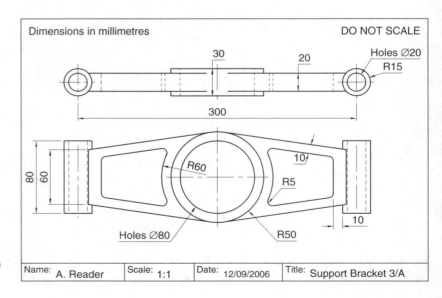

Fig. 14.8 Working drawing for the second example

7. With the **Mirror** tool, mirror the cylinders to the other end of the union.
8. Make the **Top** viewport current and with the **Move** tool, move the cylinders to their correct position at the ends of the union. Form a union between all parts on screen.
9. Make the **SE Isometric** viewport current and *enter* **render** at the command line, followed by a *click* on the **Render** button in the dialog which appears.

Fig. 14.9 shows the result.

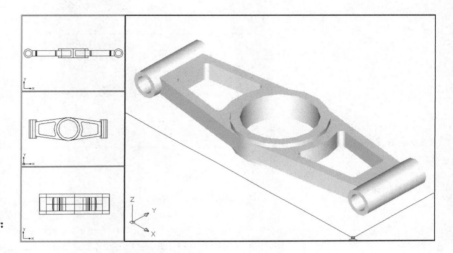

Fig. 14.9 Second example – **Four: Left viewports**

Third example – Three: Right viewports (Fig. 14.11)

1. Open the **Three: Right** viewport layout from the **Viewports** dialog. Make sure the **3D** setup is chosen.
2. *Click* in the **Top** viewport (top left-hand viewport) and change it to a **3D Views/Right** view from the **View** drop-down menu.
3. In what is now the **Right** viewport construct a pline outline to the dimensions in Fig. 14.10.

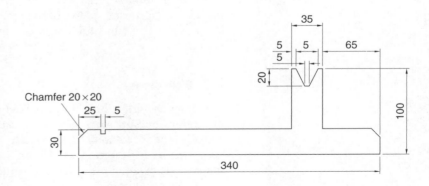

Fig. 14.10 Third example – outline for solid of revolution

4. Call the **Revolve** tool from the **Solids** toolbar and revolve the outline through 360°.
5. Make the **SE Isometric** viewport current and render the view (Fig. 14.11).

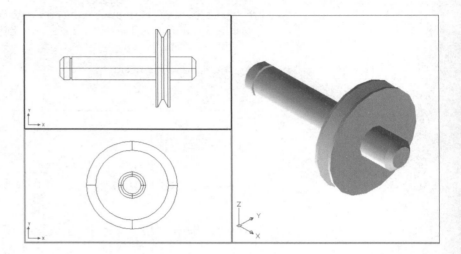

Fig. 14.11 Third example –
Three: Right viewports

Notes

1. When working in viewport layouts such as in the above three examples, it is important to make good use of the **Zoom** tool, mainly because the viewports are smaller than the whole screen viewport when working in AutoCAD 2005 as a single viewport.
2. As in all other forms of constructing drawings in AutoCAD 2005, frequent toggling of **SNAP**, **ORTHO** and **GRID** will allow speedier and more accurate working.

Revision notes

1. Outlines suitable for use when constructing 3D models can be constructed using the 2D tools such as **Line**, **Arc**, **Circle** and **Polyline**. Such outlines must be changed to either closed polylines or regions before being incorporated in 3D models.
2. The use of multiple viewports can be of value when constructing 3D models in that various views of the model appear enabling the operator to check the accuracy of the 3D appearance throughout the construction period.

Exercises

1. Using the **Cylinder**, **Box**, **Sphere**, **Wedge** and **Fillet** tools, together with the **Union** and **Subtract** tools and working to any sizes thought suitable, construct the 'head' in the **Three: Left** viewport as shown in Fig. 14.12.
2. Using the tools **Sphere**, **Box**, **Union** and **Subtract** and working to the dimensions given in Fig. 14.14, construct the 3D solid model as shown in the rendering in Fig. 14.13.

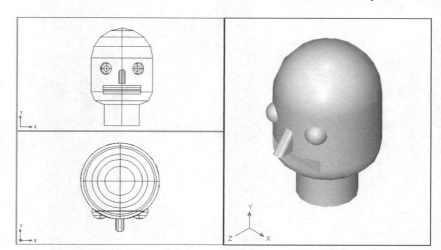

Fig. 14.12 Exercise 1

Fig. 14.13 Exercise 2

3. Each link of the chain shown in the rendering in Fig. 14.15 has been constructed using the **Extrude** tool – extruding a small circle along an elliptical path. Copies of the link were then made, half of which were rotated in a **3D Views/Right** view and then moved into their position relative to the other links. Working to suitable sizes, construct a link and from the link construct the chain as shown.

4. A two-view orthographic projection of a rotatable lever from a machine is given in Fig. 14.16 together with a rendering of the 3D model (Fig. 14.17) constructed to the given details.

  Construct the 3D model drawing in a **Four: Equal** viewport setting, and render the 3D model in the **Isometric** viewport.

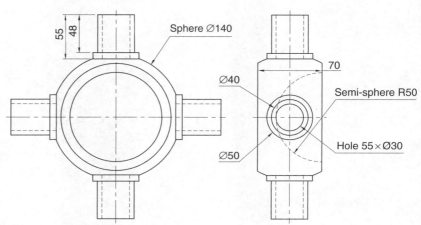

Fig. 14.14 Exercise 2 – working drawing

Fig. 14.15 Exercise 3

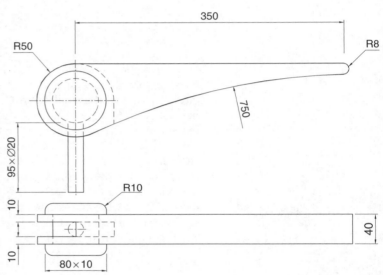

Fig. 14.16 Exercise 4 –
orthographic projection

Fig. 14.17 Exercise 4

5. Working in a **Three: Left** viewport setting, construct a 3D model of the faceplate to the dimensions given in Fig. 14.18. With the **Mirror** tool, mirror the model to obtain an opposite facing model. In the **Isometric** viewport call the **Hide** tool (Fig. 14.19).

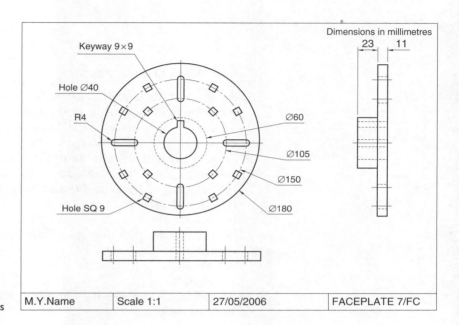

Fig. 14.18 Exercise 5 – dimensions

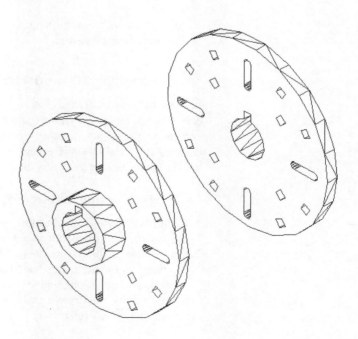

Fig. 14.19 Exercise 5

CHAPTER 15

The modification of 3D models

Aims of this chapter

1. To demonstrate how 3D models can be saved as blocks for insertion into other drawings via the **DesignCenter**.
2. To show how a library of 3D models in the form of blocks can be constructed to enable the models to be inserted into other drawings.
3. To give examples of the use of the following tools from the **Operations** sub-menu from the **Modify** drop-down menu:
   (a) **3D Array – Rectangular and Polar 3D arrays**
   (b) **Mirror 3D**
   (c) **Rotate 3D**.
4. To give examples of the use of the following tools from the **Solids** toolbar:
   (a) **Slice**
   (b) **Section**.
5. To give an example of the use of the **Mass Properties** tool from the **Inquiry** toolbar.
6. To show how to obtain different views of 3D models in 3D space using:
   (a) **3D Views** from the **View** drop-down menu.
   (b) **Viewpoint Presets**.

Creating 3D model libraries

In the same way as 2D drawings of parts such as electronics symbols, engineering parts, building symbols and the like can be saved in a file as blocks and then opened into another drawing by *dragging* the appropriate block drawing from the **DesignCenter**, so can 3D models be saved as blocks in a file for *dragging and dropping* into another drawing from the **DesignCenter**.

First example – inserting 3D blocks (Fig. 15.4)

1. Construct individual 3D models of the parts for a lathe milling wheel holder to the dimensions given in Fig. 15.1.
2. Save each of the 3D models of the parts to file names as given in the drawing as blocks using the **Make Block** tool from the **Draw** toolbar. When all seven blocks have been saved, the drawings on screen can be deleted. Save the drawing with its blocks to a suitable file name. In this example this is **Fig01.dwg**.

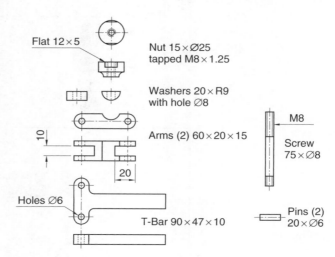

Fig. 15.1 The components of a
lathe milling wheel holder

Fig. 15.2 Calling the
DesignCenter to screen

3. Set up a **Four: Equal** viewports setting.
4. Open the **DesignCenter** with a *click* on its tool icon in the **Standard** toolbar (Fig. 15.2), or by pressing the **Ctrl** and **2** keys of the keyboard.
5. In the **DesignCenter** *click* the directory **Chapter15**, followed by another *click* on **Fig03.dwg** and yet another *click* on **Blocks**. The saved blocks appear as icons in the right-hand area of the **DesignCenter**.
6. *Drag and drop* the blocks one by one into one of the viewports on screen. Fig. 15.3 shows the **Nut** block ready to be *dragged* into position in the **Front** viewport. As the blocks are *dropped* on screen, they will need moving into their correct positions in relation to other parts of the assembly by using the **Move** tool from the **Modify** toolbar in suitable viewports.

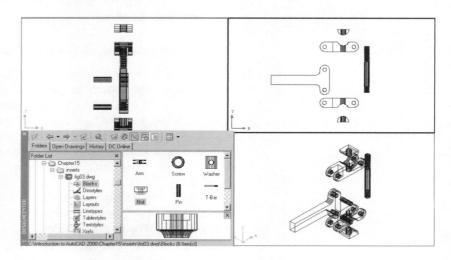

Fig. 15.3 First example –
inserting 3D blocks

7. Using the **Move** tool, move the individual 3D models into their final places on screen and render the **SE Isometric** viewport (Fig. 15.4).

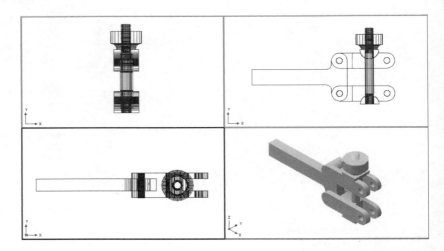

Fig. 15.4 First example –
inserting 3D blocks

Notes

1. It does not matter which of the four viewports any one of the blocks is *dragged and dropped* into. The part automatically assumes the view of the viewport.
2. If a block destined for layer **0** is *dragged and dropped* into the layer **Centre** (which in our **acadiso.dwt** is of colour **red** and of linetype **CENTER2**), the block will take on the colour (red) and linetype of that layer (**CENTER2**).
4. As a block is being *dragged and dropped* from the **DesignCenter**, the command line shows:

   Command: DROPGEOM GROUP

5. In this example, the blocks are 3D models and there is no need to use the **Explode** tool option.

Second example – a library of fastenings (Fig. 15.6)

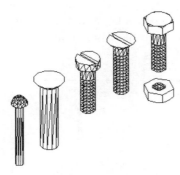

Fig. 15.5 Second example – the five fastenings

1. Construct a number of engineering fastenings. The number constructed does not matter. In this example only five have been constructed – a 10 mm round head rivet, a 20 mm countersunk head rivet, a cheese head bolt, a countersunk head bolt and a hexagonal head bolt together with its nut (Fig. 15.5). With the **Make Block** tool save each separately as a block, erase the original drawings and save the file to a suitable file name – in this example this is **Fig05.dwg**.
2. Open the **DesignCenter**, *click* on the **Chapter15** directory, followed by a *click* on **Fig05.dwg**. Then *click* again on **Blocks** in the content list of **Fig05.dwg**. The five 3D models of fastenings appear as icons in the right-hand side of the **DesignCenter** (Fig. 15.6).
3. Such engineering fastenings can be *dragged and dropped* into position in any engineering drawing where the fastenings are to be included.

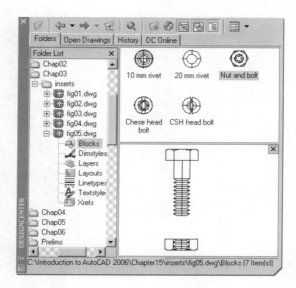

Fig. 15.6 Second example –
a library of fastenings

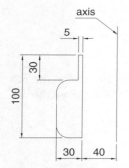

Fig. 15.7 Example of constructing
a 3D model – outline for solid of
revolution

An example of constructing a 3D model (Fig. 15.9)

A three-view projection of a pressure head is shown in Fig. 15.8. To construct a 3D model of the head:

1. From **3D Views** select the **Front** view.
2. Construct the outline to be formed into a solid of revolution (Fig. 15.7) and with the **Revolve** tool, produce the 3D model of the outline.
3. Place the screen in the **3D Views/Top** view and with the **Cylinder** tool, construct cylinders as follows:
   (a) In the centre of the solid already constructed – radius **50** and height **50**.
   (b) With the same centre – radius **40** and height **40**. Subtract this cylinder from that of radius **50**.
   (c) At the correct centre – radius **10** and height **25**.
   (d) At the same centre – radius **5** and height **25**. Subtract this cylinder from that of radius **10**.
4. With the **Array** tool, form a **6** times polar array of the last two cylinders based on the centre of the 3D model.
5. Place the drawing in the **Front** view.
6. With the **Move** tool, move the array and the other two cylinders to their correct positions relative to the solid of revolution so far formed.
7. With the **Union** tool form a union of the array and other two solids.
8. Place the screen in the **3D Views/Right** view.
9. Construct a cylinder of radius **30** and height **25** and another of radius **25** and height **60** central to the lower part of the 3D solid so far formed.
10. Place the screen in the **3D Views/Top** view and with the **Move** tool move the two cylinders into their correct position relative to the 3D solid.

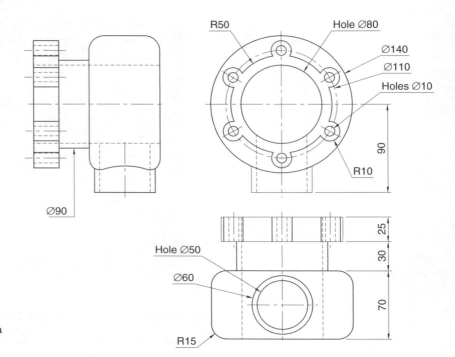

Fig. 15.8 Orthographic drawing for the example of constructing a 3D model

Fig. 15.9 Example of constructing a 3D model

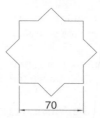

Fig. 15.10 Example – 3D Array – the star pline

11. With **Union**, form a union between the radius **30** cylinder and the 3D model and with **Subtract**, subtract the radius **25** cylinder from the 3D model.

12. Call **Render** and render the model (Fig. 15.9).

Note

This 3D model could equally as well have been constructed in a three or four viewports setting.

The 3D Array tool

First example – Rectangular array (Fig. 15.12)

1. Construct the star-shaped pline (Fig. 15.10) and extrude it to a height of **20**.

2. *Click* on **Modify** in the menu bar and in the drop-down menu which appears *click* on **3D Operation**, followed by another *click* on **3D Array** in the sub-menu which appears (Fig. 15.11). The command line shows:

Command:_3darray
Select objects: *pick* the extrusion **1 found**
Select objects: *right-click*
Enter the type of array [Rectangular/Polar] <R>: *right-click*
Enter the number of rows (—) <1>: *enter* **3** *right-click*
Enter the number of columns (III): *enter* **3** *right-click*
Enter the number of levels (. . .): *enter* **4** *right-click*

Specify the distance between rows (—): 50
Specify the distance between columns (III): 50
Specify the distance between levels (. . .): 150
Command:

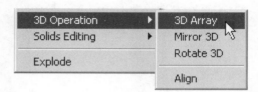

Fig. 15.11 Selecting **3D Array** from the **Modify** drop-down menu

3. Place the screen in the **3D Views/SW Isometric** view.
4. Call the **Hide** tool.

Second example – Polar array (Fig. 15.13)

1. Use the same star-shaped 3D model.
2. Call the **3D Array** tool again. The command line shows:

Command:_3darray
Select objects: *pick* the extrusion **1 found**
Select objects: *right-click*
Enter the type of array [Rectangular/Polar] <R>: *enter* p (Polar)
    right-click
Enter number of items in the array: 12
Specify the angle to fill (+ = ccw, − = cw) <360>: *right-click*
Rotate arrayed objects? [Yes/No] <Y>: *right-click*
Specify center point of array: 235,125
Specify second point on axis of rotation: 300,200
Command:

3. Place the screen in the **3D Views/SW Isometric** view.
4. Render the drawing.

Third example – Polar array (Fig. 15.16)

1. Construct a solid of revolution in the form of an arrow to the dimensions as shown in Fig. 15.14.
2. Make a **12** times polar array around a central point (Fig. 15.15).
3. Call the **Union** tool and form a union of all 12 items in the array.
4. Place the screen in the **3D Views/Front** view.
5. With the **Copy Object** tool, copy the union of the 12 arrows in the array as follows:

Command:_copy
Select objects: *click* at any point on one of the arrows in the union – the whole union is *picked*
Specify base of displacement: *pick* any point in the union
Specify second point of displacement: *pick* the same point – the copy is on top of the original
Command:

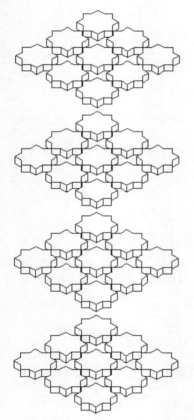

Fig. 15.12 First example – **Rectangular Array**

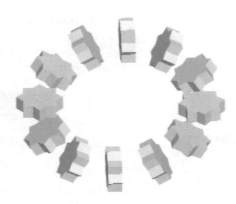

Fig. 15.13 Second example –
Polar array

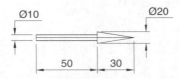

Fig. 15.14 Third example – **Polar array** – the 3D model to be arrayed

6. Call the **Rotate** tool from the **Modify** drop-down menu. The command line shows:

Command:_rotate
Select objects: *enter* l (for Last) *right-click*
Select objects: *right-click*
Specify base point: *pick* the centre of the array
Specify rotation angle: *enter* **90** *right-click*
Command:

7. Repeat step **6** to rotate the original array through **45°** and also through **135°**.
8. Place the composite array in the **3D Views/SW Isometric** view and render.

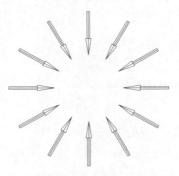

Fig. 15.15 Third example – **Polar array** – the first array

Fig. 15.16 Third example – **Polar array**

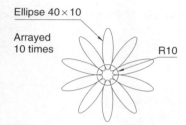

Ellipse 40 × 10

Arrayed
10 times

R10

Fig. 15.17 First example –
Mirror 3D – plan of object to
be mirrored

Note

In this example the array did not require the use of the **3D Array** tool, because all arrays were on objects in the same plane.

The Mirror 3D tool

First example – Mirror 3D (Fig. 15.20)

1. Construct the array of ellipses and circle as shown in Fig. 15.17.
2. Change the 11 objects into regions, form a union of the ellipses and subtract the circle from the ellipses.
3. Extrude the region to a height of **5** and render. The rendering in the **SW Isometric** view is shown in Fig. 15.18.
4. *Click* on **Mirror 3D** in the **3D Operation** sub-menu of the **Modify** drop-down menu (Fig. 15.19). The command line shows:

Fig. 15.18 First example –
Mirror 3D – the object to be
mirrored

Command:_mirror3d
Select objects: *pick* the extrusion **1 found**
Select objects: *right-click*
Specify first point on mirror plane (3 points): 80,130,100
Specify second point on mirror plane: 170,220,50
Specify third point on mirror plane: 180,160,20
Delete source objects? [Yes/No] <N>: *right-click*
Command:

Fig. 15.19 Calling **Mirror 3D** from
the **Modify** drop-down menu

Fig. 15.20 First example –
Mirror 3D

Fig. 15.21 Second example –
Mirror 3D – the 3D model

Second example – Mirror 3D (Fig. 15.23)

1. Construct a solid of revolution in the shape of a bowl in the **3D Views/Front** view (Fig. 15.21).
2. *Click* **Mirror 3D** in the **3D Operations** sub-menu of the **Modify** drop-down menu. The command line shows:

Command:_mirror3d
Select objects: *pick* the bowl **1 found**
Select objects: *right-click*
Specify first point on mirror plane (3 points): *pick*
Specify second point on mirror plane: *pick*
Specify third point on mirror plane: *enter* **.xy** *right-click* **(need Z):**
 enter **1** *right-click*
Delete source objects? [Yes/No] <N>: *right-click*
Command:

Note

The line in the illustration shows the top edge of an imaginary plane vertical to the plane on which the drawing is being constructed (Fig. 15.22).

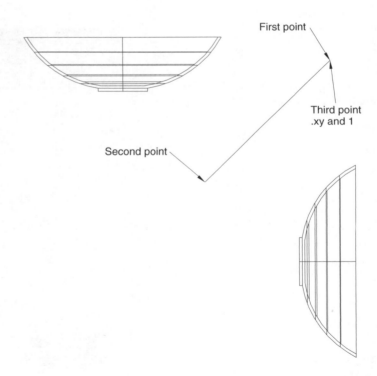

First point

Third point
.xy and 1

Second point

Fig. 15.22 Second example –
Mirror 3D – the result in a
front view

3. Place in the **3D Views/SW Isometric** view.
4. Call **Render**. The result is shown in Fig. 15.23.

Fig. 15.23 Second example –
Mirror 3D

The Rotate 3D tool

Example – Rotate 3D (Fig. 15.25)

1. Using the same 3D model of a bowl as for the last example, but treating the line as a line and not as an imaginary plane (Fig. 15.24), call the **Rotate 3D** tool from the **3D Operations** sub-menu of the **Modify** drop-down menu. The command line shows:

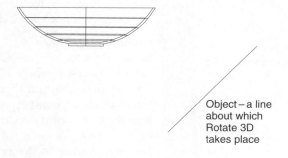

Object – a line
about which
Rotate 3D
takes place

Fig. 15.24 Example – **Rotate 3D** –
the bowl and the line of rotation

Command:_rotate3d
Select objects: *pick* the bowl
Select objects: *right-click*
Specify first point on axis or define by
[Object/Last/View/Xaxis/Yaxis/Zaxis/2points]: *enter* **o** (Object)
    right-click
Specify a line, circle, arc or 2D-polyline segment: *pick* the line
Specify rotation angle: *enter* **60** *right-click*
Command:

Fig. 15.25 Example – **Rotate 3D**

2. Call **Render** and render the rotated bowl (Fig. 15.25).

The Slice tool

First example – Slice (Fig. 15.29)

1. Construct a 3D model of the rod link device shown in the two-view projection (Fig. 15.26).

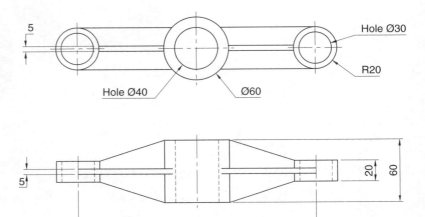

Fig. 15.26 First example – **Slice** – the two-view drawing

Fig. 15.27 The **Slice** tool icon from the **Solids** toolbar

2. Place the 3D model in the **3D Views/Top** view.
3. *Click* the **Slice** tool icon in the **Solids** toolbar (Fig. 15.27). The command line shows (Fig. 15.28 shows the *picked* points):

Command:_slice
Select objects: *pick* the 3D model
Select objects: *right-click*
Specify first point on slicing plane <3points>: *pick*
Specify second point on slicing plane <3points>: *pick*
Specify third point on slicing plane <3points>: *enter* **.xy** *right-click*
of *pick* first point again (**need Z**): *enter* **1** *right-click*
Specify a point on desired side of the plane or [keep Both sides]: *enter* **b** (Both) *right-click*
Command:

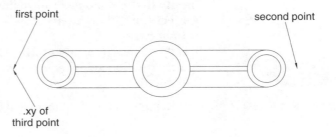

Fig. 15.28 First example – **Slice** – the *pick* points

Fig. 15.29 First example – **Slice**

4. With the **Move** tool, move the lower half of the sliced model away from the upper half.
5. Place the 3D model(s) in a **SW Isometric** view.
6. Call **Render**.

The result is shown in Fig. 15.29.

Second example – Slice (Fig. 15.30)

1. Construct the closed pline (Fig. 15.30) and with the **Revolve** tool, form a solid of revolution from the pline.

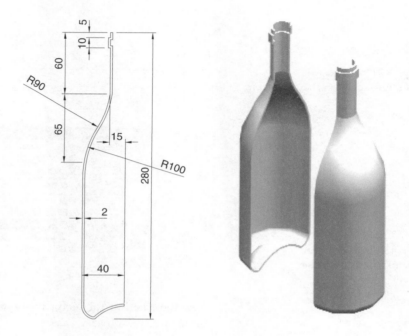

Fig. 15.30 Second example – **Slice**

2. With the **Slice** tool and working to the same sequence as for the first **Slice** example, form two halves of the 3D model and render.

The Section tool

First example – Section (Fig. 15.34)

1. Construct a 3D model to the information given in Fig. 15.31 on layer **0**. Note there are three objects in the model – a box, a lid and a cap.
2. Place the model in the **3D Views/Top** view.
3. *Click* in the layer field of the **Layers** toolbar (Fig. 15.32) and *click* again on **Construction** to make it the current layer. Its colour is cyan.

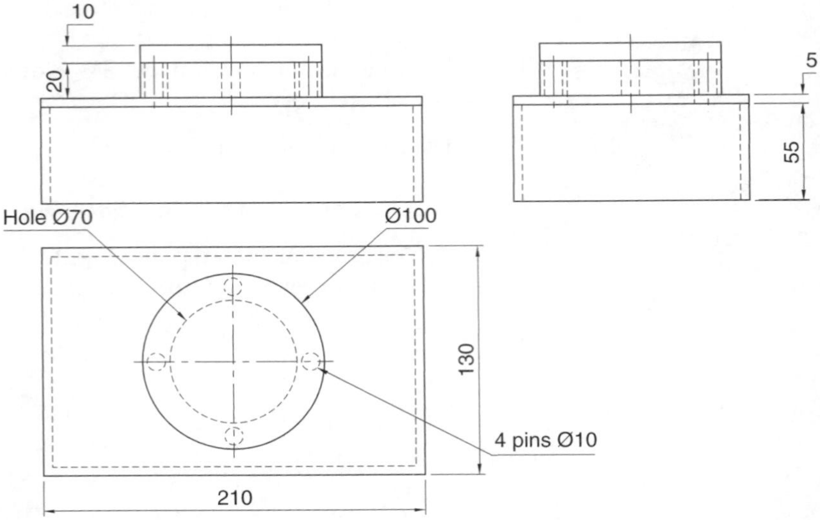

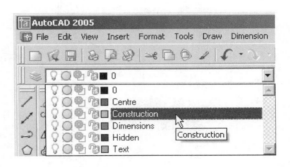

Fig. 15.31 First example –
Section – orthographic projection

Fig. 15.32 Making layer
Construction current

Fig. 15.33 The **Section** tool icon
from the **Solids** toolbar

4. *Click* the **Section** tool icon in the **Solids** toolbar (Fig. 15.33). The command line shows:

Command:_section
Select objects: window the 3D model **3 found**
Select objects: *right-click*
Specify first point on section plane <3points>: *pick*
Specify second point on plane: *pick*
Specify third point on plane: *enter* **.xy** *right-click*
of *pick* first point **(need Z):** *enter* **1** *right-click*
Command:

And a cyan line showing the top edge of the section plane appears in the view.

> *Note*

The three points *picked* above are similar to those given in the previous examples of using the **Slice** tool.

5. Turn **Layer 0** (on which the 3D model was constructed) off, leaving only the cyan line showing on screen.
6. Place the screen in the **3D Views/Front** view and **Zoom** to **1**. The outlines of the section appear.

   Note

   The sectional view as given in the outlines (Fig. 15.34) will not be a correct view as required in general engineering drawing practice for the following reasons:

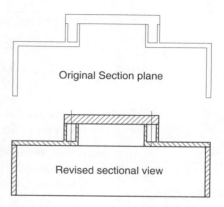

Original Section plane

Revised sectional view

Fig. 15.34 First example – **Section**

(a) There are no hatch lines. This is usually acceptable because in some circumstances hatching of sections is not expected, but in this example hatch lines are to be included.
(b) In engineering sectional views, parts such as the pins holding the cap onto the lid would usually be shown by outside views in a sectional view.

7. Amend the drawing by adding lines as necessary and hatching using the **ANSI31** hatch pattern as shown in the lower drawing of Fig. 15.34.

Second example – Section (Fig. 15.36)

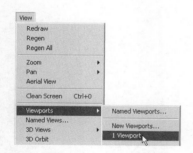

Fig. 15.35 Calling **1 Viewport** from the **View** drop-down menu

1. Open the drawing of the lathe tool holder constructed in answer to the first example in this chapter (Fig. 15.4). The drawing is in a **Four: Equal** viewports setting. *Click* in the **Top** viewport, and from the **View** drop-down menu *click* **1 Viewport** in the **Viewports** sub-menu (Fig. 15.35). The assembly appears in a full size single viewport.
2. Call the **Section** tool from the **Solids** toolbar and by turning **Layer 0** off and placing the 3D model in the **3D Views/Front** view, the section plane used when the 3D model was acted upon by the **Section** tool working in the **Construction** layer shows on screen (Fig. 15.36).

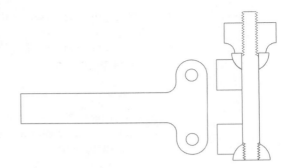

Fig. 15.36 Second example –
Section

Region/Mass Properties

Fig. 15.37 The **Mass Properties**
tool icon in the **Inquiry** toolbar

The Massprop tool

Example – Massprop (Fig. 15.40)

1. Open the 3D model of the rod link device shown in Fig. 15.26 on page 208.
2. Call the **Mass Properties** tool – either with a *click* on its tool icon in the **Inquiry** toolbar (Fig. 15.37) or by *entering* **massprop** at the command line. When called, the command line shows:

Command:_massprop
Select objects: *pick* the 3D model **1 found**
Select objects: *right-click*

An AutoCAD Text Window appears showing the mass properties of the *picked* 3D model (Fig. 15.38).

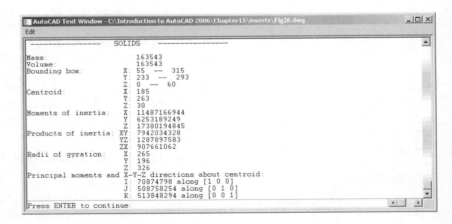

Fig. 15.38 **Massprop** tool. The
AutoCAD Text window showing
the mass properties

Press ENTER to continue:
Write analysis to a field [Yes/No]: *enter* **y** *right-click*

And the **Create Mass and Area Properties File** dialog appears showing the name of the file, **Fig26_mass.mpr** (Fig. 15.39). *Click* the **Save** button of the dialog and the file is saved – in this case to the file name **Fig26_mass.mpr**.

Command:

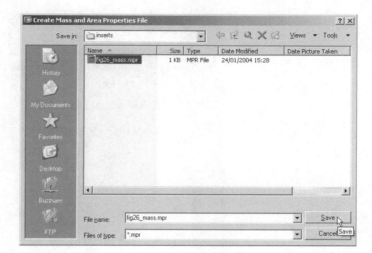

Fig. 15.39 **Massprop** tool. The **Create Mass and Area Properties File** dialog

3. Open the application **Window Excel** (a database program). *Click* on the name **File** in its menu bar, followed by another *click* on **Open** . . . and from the **Open** sub-menu which then appears, select the file **Fig26_mass.mpr**. The details from the AutoCAD Text Window appear as a database in Excel (Fig. 15.40).

	A	B	C	D
1				
2		SOLIDS		
3				
4	Mass:		163543	
5	Volume:		163543	
6	Bounding box:	X:	55 -- 315	
7		Y:	233 -- 293	
8		Z:	0 -- 60	
9	Centroid:	X:	185	
10		Y:	263	
11		Z:	30	
12	Moments of inertia:	X:	11487166944	
13		Y:	6253189249	
14		Z:	17380194845	
15	Products of inertia:	XY:	7942034328	
16		YZ:	1287897583	
17		ZX:	907661062	
18	Radii of gyration:	X:	265	
19		Y:	196	
20		Z:	326	
21	Principla moments and	X-Y-Z	directions	about centroid
22		I:	70874798	along [1,0,0]
23		J:	508758254	along [0,1,0]
24		K:	5138488284	along [0,0,1]

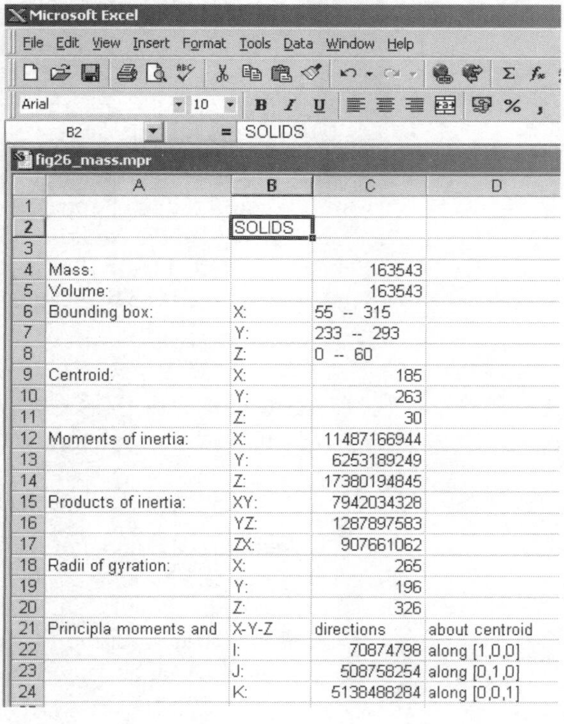

Fig. 15.40 Part of the **Excel** database file

4. When closing the **Excel** program a warning window appears informing the operator as to how to save an ***.mpr** file as an Excel file, Follow the instruction to save the file in Excel format (***.xls**).
5. The file can now be sent to another person as an attachment in an email.

Views of 3D models

Fig. 15.41 is a two-view projection of a model of an arrow.

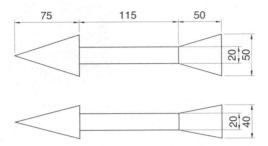

Fig. 15.41 Two views of the arrow

Some of the possible viewing positions of a 3D model which can be obtained by using the **3D Views** positions have already been used in this book. The views in Fig. 15.42 show all of the viewing positions of the 3D model of the arrow using the viewing positions from the **3D Views** sub-menu of the **Views** drop-down menu.

The Viewpoint Presets dialog

There are other methods of obtaining a variety of viewing positions of a 3D model. One method is by using the **UCS** (User Coordinate System) which will be described in a later chapter. Another method is by using the **Viewpoint Presets** dialog called with a *click* on **Viewpoint Presets** . . . in the **3D Views** sub-menu of the **View** drop-down menu (Fig. 15.43).

When the dialog appears with a 3D model on screen, *entering* figures for degrees in the **X Axis** and **XY Plane** fields, followed by a *click* on the dialog's **OK** button, causes the model to take up the viewing position indicated by these two angles.

Note

The **Relative to UCS** radio button must be checked on to allow the 3D model to position along the two angles.

Example – Viewpoint Presets

1. With the 3D model of the arrow on screen, *click* **Viewpoint Presets** . . . in the **3D Views** sub-menu of the **View** drop-down menu. The dialog appears.
2. *Enter* **30** in both the **From X Axis** and the **From XY Plane** fields and *click* the **OK** button of the dialog.

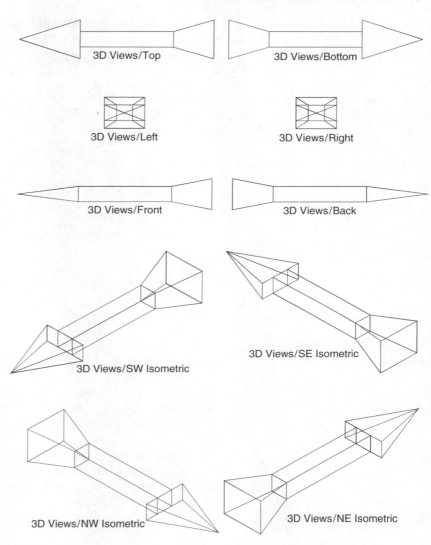

Fig. 15.42 The views from **3D Views**

Fig. 15.43 Calling the **Viewpoint Presets** dialog

3. The 3D model takes up the viewing position indicated by the two angles (Fig. 15.44).

Revision notes

1. 3D models can be saved as blocks in a similar manner to the method of saving 2D drawings as blocks.
2. Libraries can be made up from 3D model drawings.
3. 3D models saved as blocks can be inserted into other drawings via the **DesignCenter**.
4. Arrays of 3D model drawings can be constructed in 3D space using the **3D Array** tool.
5. 3D models can be mirrored in 3D space using the **Mirror 3D** tool.

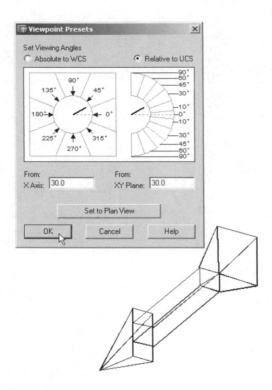

Fig. 15.44 Example – **Viewpoint Presets**

6. 3D models can be rotated in 3D space using the **Rotate 3D** tool.
7. 3D models can be cut into parts with the **Slice** tool.
8. Sectional views can be obtained from 3D model using the **Section** tool.
9. The properties and area of a 3D model can be obtained in database format using the **Mass Properties** tool.
10. Both **3D Views** viewing positions and **Viewpoint Presets** can be used for the placing of 3D models in different viewing positions in 3D space.

Exercises

1. Fig. 15.45 shows a rendering of the 3D model for this exercise. Fig. 15.46 is a three-view projection of the model. Working to the details given in Fig. 15.46, construct the 3D model.
2. Construct a 3D model drawing of the separating link shown in the two-view projection (Fig. 15.47). With the **Slice** tool, slice the model into two parts and remove the rear part.

   Add suitable lighting and a material to the half view. Place the front half in a suitable isometric view from the **3D Views** sub-menu. Render the resulting model.
3. Working to the dimensions given in the three orthographic projections (Fig. 15.48), construct an assembled 3D model of the one part inside the other.

Fig. 15.45 Exercise 1 – a rendering

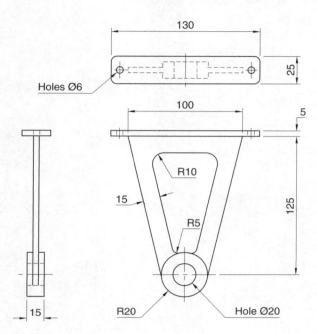

Fig. 15.46 Exercise 1 – three-view projection

With the **Slice** tool, slice the resulting 3D model into two equal parts, place in an isometric view and call the **Hide** tool as indicated in Fig. 15.49.

4. Construct a solid of revolution of the jug shown in the orthographic projection (Fig. 15.50). Construct a handle from an extrusion of a circle along a semicircular path. Union the two parts.

Place the resulting model in the **3D Views/Plan** view and add lighting. Also add a suitable glass material.

Place the 3D model in a suitable isometric view and render.

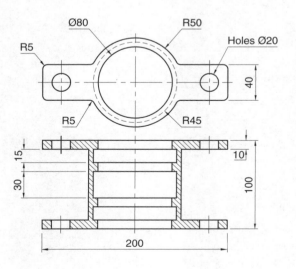

Fig. 15.47 Exercise 2

Fig. 15.48 Exercise 3 –
orthographic projections

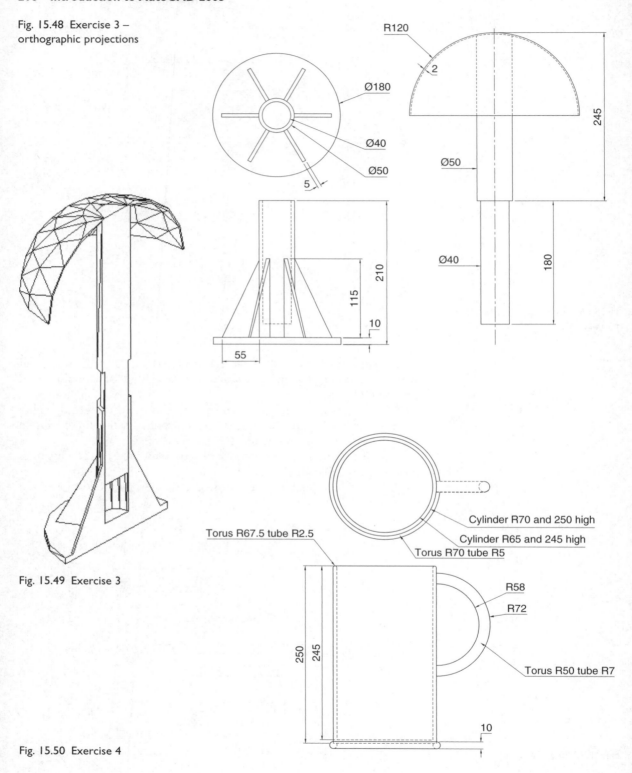

R120

2

Ø180

Ø40

Ø50

5

245

Ø50

Ø40

180

210

115

10

55

Fig. 15.49 Exercise 3

Cylinder R70 and 250 high

Cylinder R65 and 245 high

Torus R70 tube R5

Torus R67.5 tube R2.5

R58

R72

Torus R50 tube R7

250

245

10

Fig. 15.50 Exercise 4

CHAPTER 16

Rendering

Aims of this chapter

1. To introduce the use of the **Render** tools in producing photographic like images of 3D solid models.
2. To show how to illuminate a 3D solid model to obtain good lighting effects when rendering.
3. To give examples of the rendering of 3D solid models.
4. To introduce the idea of adding materials to 3D solid models in order to obtain a realistic appearance to a rendering.
5. To demonstrate the use of the forms of shading available using the **3D Orbit** tool.
6. To demonstrate methods of printing rendered 3D solid models.

The Render tools

The use of the **Render** tool has already been shown in examples in previous pages. More realistic images can be produced using the tool, if lights, colours and/or materials are added to the 3D model being rendered. All the tools in the **Render** toolbar have already been shown on page 165.

Render lights

There are four types of lights available when using AutoCAD – **Ambient** lights, **Point** lights, **Distant** lights and **Spotlights**.

Ambient lighting is taken as the general overall light that is all around and surrounding any object. When rendering in AutoCAD **Ambient** light is generally set at a default value of **0.3**.

Point lights shed light in all directions from the position in which the light is placed and diminish in intensity the further from the light that an object is placed.

Distant lights send parallel rays of light from their position in the direction chosen by the operator. There is no diminution of the light intensity no matter how far from an object the lights are placed.

Spotlights illuminate as if from a spotlight. The light is in a direction set by the operator and is in the form of a cone, with a 'hotspot' cone giving a brighter spot on the model being lit.

Placing lights to illuminate a 3D model

Any number of the three types of lights – **Point**, **Distant** and **Spotlight** can be positioned in 3D space as wished by the operator.

In general reasonably good lighting effects can be obtained by placing a **Point** light high above the object(s) being illuminated, with a **Distant** light placed pointing towards the object at a distance from the front and above the general height of the object(s) and with a second **Distant** light pointing towards the object(s) from one side and not as high as the first **Distant** light. If desired **Spotlights** can be used either on their own or in conjunction with the other two forms of lighting. **Ambient** light is automatically placed in the 3D space occupied by the model and other lighting when any form of lighting is added to a scene.

First example – rendering a 3D model (Fig. 16.5)

1. Construct a 3D model of the wing nut shown in the two-view projection in Fig. 16.1.

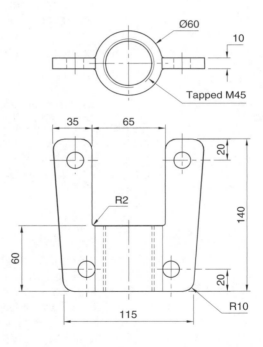

Fig. 16.1 First example – **Rendering** – two-view projection

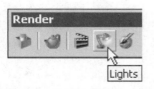

Fig. 16.2 The **Lights** tool icon in the **Render** toolbar

2. Place the 3D model in the **3D Views/Top** view, **Zoom** to **1** and with the **Move** tool, move the model to the upper part of the AutoCAD drawing area.
3. *Click* on the **Lights** tool icon in the **Render** toolbar (Fig. 16.2). The **Lights** dialog appears (Fig. 16.3).
4. In the dialog *click* on the **New** . . . button to the left of the field reading **Point Light**. The dialog is replaced by the **New Point Light** dialog (Fig. 16.4).

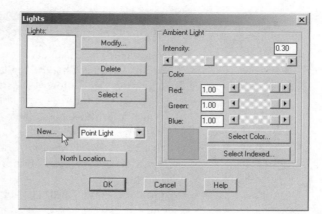

Fig. 16.3 First example –
Rendering – the **Lights** dialog

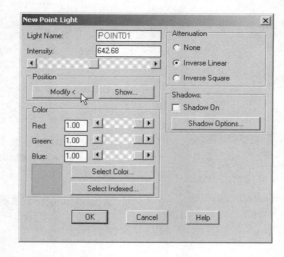

Fig. 16.4 First example –
Rendering – the **New Point
Light** dialog

5. In the **New Point Light** dialog *enter* a name in the **Light Name** field –
in this example this is **POINT01**. Then *click* on the **Modify<** button.
The dialog disappears and the command line shows:

   Command:_LIGHT
   Enter light location <current>: *enter* **.xy** *right-click*
   of *click* in the centre of the plan of the 3D model
   (need Z): *enter* **500** *right-click*

   The dialog reappears. *Click* its **OK** button. The **Lights** dialog reappears.
6. In the dialog, *click* the arrow to the right of the **Point Light** field and
from the popup list which appears *click* **Distant Light**, followed by a
click on the **New** . . . button. The **Distant Lights** dialog appears. In
the **Light Name** field *enter* **DIST01**, followed by a *click* on the
Modify button. The dialog disappears and the command line shows:

   Enter direction TO <current>: *enter* **.xy** *right-click*
   of *pick* the centre of the plan **(need Z):** *enter* **70** (halfway up the 3D
      model) *right-click*

Enter light direction FROM <current>: *enter* **.xy** *right-click* **of** *enter* **150,−20** *right-click* (**need Z**): *enter* **500** *right-click*

And the dialog reappears.

7. Modify another **Distant** Light to the same position from −80,−20,300.
8. Click the **OK** button of the dialog. It is replaced by the **Lights** dialog. *Click* its **OK** button.
9. Place the model in the **3D View/SW Isometric** view, **Zoom** window the drawing to a suitable size and *click* the **Render** button in the **Render** toolbar. In the dialog which appears *click* the **Render** button. The model renders, but will probably be far too brightly lit.
10. Go back to the **Lights** dialog and select any one of the light names in the **Lights** list and in the appropriate light dialog, alter the **Intensity** of the light by moving the **Intensity** slider by *dragging* with the mouse. It will probably be necessary to lessen the light intensity of each of the three lights in this manner.
11. Now render again. If necessary again alter the light intensity of the lights until satisfied.

The resulting rendering is shown in Fig. 16.5.

Fig. 16.5 First example –
Rendering

Adding materials to a model

1. *Click* on the **Materials Library** tool icon in the **Render** toolbar (Fig. 16.6). The **Materials Library** dialog appears (Fig. 16.7).

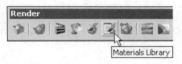

Fig. 16.6 The **Materials Library**
tool icon in the **Render** toolbar

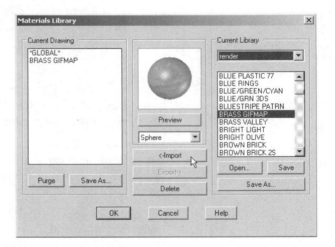

Fig. 16.7 **Rendering** – the
Materials Library dialog

2. In the dialog *click* **BRASS GIFMAP** in the **Current Library** list, followed by a *click* on the **Preview** button.
3. If satisfied with the preview, *click* the **Import** button. The name **BRASS GIFMAP** appears in the **Current Drawing** list.
4. *Click* the dialog's **OK** button.

5. *Click* on the **Materials** tool icon in the **Render** toolbar. The **Materials** dialog appears showing **BRASS GIFMAP** highlighted in its **Materials** list. *Click* the dialog's **Attach** button. The dialog disappears. *Click* on the 3D model to attach the brass material to the model and when the dialog reappears, *click* its **OK** button.

6. The 3D model can now be rendered again. It will be seen that the model appears as if made in brass.

Notes

1. In each of the **Lights** dialogs there is a check box by which **Shadows** can be switched on. However note that the check box under **Rendering Options** in the **Render** dialog must also be on.

2. The limited descriptions of rendering given in these pages do not show the full value of different types of lights, materials and rendering methods. The reader is advised to experiment with the facilities available for rendering.

Second example – Rendering a 3D model (Fig. 16.9)

1. Construct 3D models of the two parts of the stand and support given in the projections in Fig. 16.8 with the two parts assembled together.

Fig. 16.8 Second example – **Rendering** – projections of the two parts

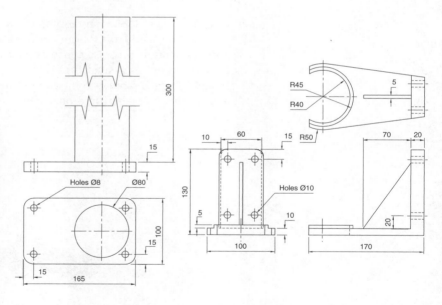

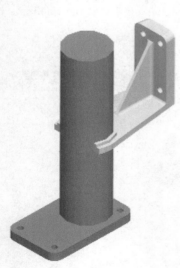

Fig. 16.9 Second example – **Rendering**

2. Place the scene in the **3D Views/Top** view and add lighting.

3. Add the two materials **BRASS VALLEY** and **METAL CHERRY RED** to the parts of the assembly and render the result.

Fig. 16.9 shows the resulting rendering.

The 3D Orbit toolbar

Bring the **3D Orbit** toolbar on screen from the toolbar menu (*right-click* in any toolbar on screen) – Fig. 16.10. Only one of the tools will be shown in an example in this book – the **3D Orbit** tool, but it is advisable to practise using other tools in the toolbar to understand how 3D models can be manipulated by using the tools in this toolbar.

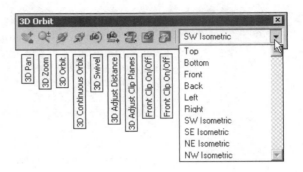

Fig. 16.10 The **3D Orbit** toolbar

Example – 3D Orbit (Fig. 16.11)

This is another tool for the manipulation of 3D models into different positions within 3D space.

1. Open the file of the second example of rendering (Fig. 16.9).
2. *Click* on the **3D Orbit** tool in the **3D Orbit** toolbar. A coloured circle appears central to the screen with small circles at its quadrants. The cursor appears as a small black circle (if outside the main circle), as an icon formed from two crossing ellipses (if inside the main circle) and as an ellipse if inside one of the smaller circles at the quadrant (Fig. 16.11).

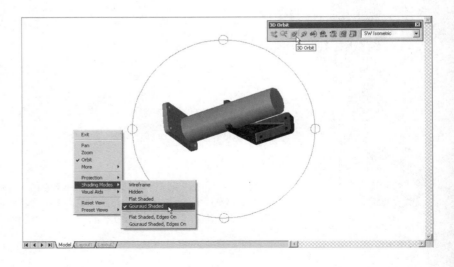

Fig. 16.11 Example – **3D Orbit**

3. With the cursor outside the circle move the mouse. The 3D model rotates within the circle.
4. With the cursor inside the circle move the mouse. The 3D model rotates around the screen.
5. With the cursor inside any one of the small quadrant circles, the 3D model can be moved vertically or horizontally as the mouse is moved.
6. *Right-click* anywhere in the screen. A *right-click* menu appears. *Click* on **Shading Modes** and again on **Gouraud Shaded** in the sub-menu which appears. The parts of the 3D model are shaded in the colours of the layers on which they had been constructed.
7. Fit the 3D model into a **Four: Equal** viewports setting. Note that the **Gouraud Shaded** mode still shows in each of the four viewports (Fig.16.12).

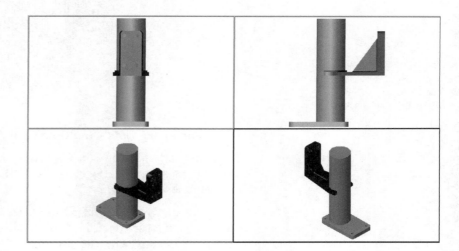

Fig. 16.12 A 3D model drawing which has been **Gouraud Shaded**

Note

The **Gouraud** shading is coloured according to the colours in which the parts of a 3D model have been constructed.

Producing hardcopy

Printing or plotting a drawing on screen using AutoCAD 2005 can be carried out from either **Model Space** or **Paper Space**. In versions of AutoCAD before AutoCAD 2004, it was necessary to print or plot from **Pspace**.

First example – printing a single copy (Fig. 16.15)

1. With a drawing to be printed or plotted on screen, *click* **Plot** . . . in the **File** drop-down menu (Fig. 16.13).
2. The **Plot** dialog appears. Set the **Plot Device** to a printer or plotter currently attached to the computer and the **Plot Settings** to a paper size to which the printer/plotter is set.

Fig. 16.13 Selecting **Plot** . . . from the **File** drop-down menu

Fig. 16.14 The *right-click* menu in
the print **Preview** window

3. *Click* the **Preview** button of the dialog and if the preview is OK, *right-click* and in the *right-click* menu which appears, *click* **Plot** (Fig. 16.14). The drawing plots producing the necessary 'hardcopy' (Fig. 16.15).

Fig. 16.15 First example –
printing a single copy

Second example – multiple view copy (Fig. 16.16)

A 3D model to be printed is a **Gouraud Shaded** 3D model which has been constructed on three layers – **Red**, **Blue** and **Green** in colour. To print a multiple view copy:

1. Place the drawing in a **Four: Equal** viewport setting.
2. Make a new layer **vports** of colour cyan and make it the current layer.
3. *Click* the **Layout1** tab. The drawing appears in **Pspace** with the **Page Setup – Layout1** dialog on top.
4. Make sure the **Plot Device** and **Plot Settings** are correct and *click* the **OK** button of the dialog.
5. A view of the 3D model appears within a cyan coloured viewport. With the **Erase** tool erase the viewport. The drawing is erased.
6. At the command line:

   Command: *enter* **mv** (MVIEW) *right-click*
   Specify corner of viewport or [ON/OFF/Fit/Shadeplot/Lock/ Object/Polygonal/Restore/2/3/4] <Fit>: *enter* **4** *right-click*
   Specify first corner or [Fit] <Fit>: *right-click*

   And a four-viewport layout appears in **Paper Space** (Fig. 16.17).
7. *Click* the **PAPER** button in the status bar. It changes to **MODEL**.
8. Set the viewports to suitable **3D View** settings.
9. *Click* **Plot** . . . in the **File** drop-down menu and when the **Plot** dialog appears, *click* its **Full Preview** button.
10. If the preview is satisfactory (Fig. 16.17), *right-click* and from the *right-click* menu *click* **Plot**. The drawing plots to produce the required four-viewport hardcopy.

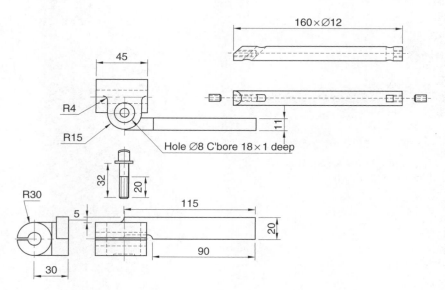

Fig. 16.16 Second example –
multiple view copy

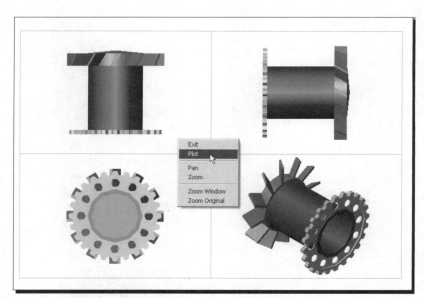

Fig. 16.17 Second example –
multiple view copy – a print
Preview

Other forms of hardcopy

When working in AutoCAD 2005, several different forms of hardcopy can
be printed or plotted determined by the settings in the **3D Orbit/Shading
Modes** settings (Fig. 16.18). As an example a single view plot review of
the same 3D model is shown in the **Hidden** shading form (Fig. 16.19).

Saving and opening 3D model drawings

3D model drawings are saved and/or opened in the same way as are 2D drawings. To save a drawing *click* **Save As** . . . in the **File** drop-down menu and in

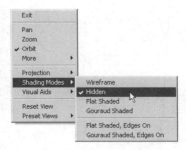

Fig. 16.18 The **3D Orbit/Shading Modes** settings menu

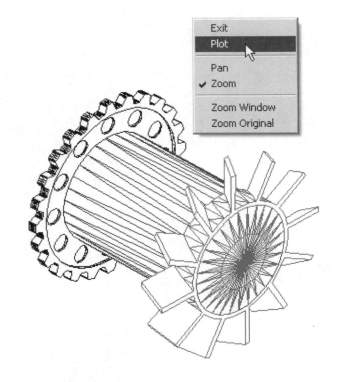

Fig. 16.19 An example of **Hidden** shading plot **Preview**

the **Save Drawing As** dialog which appears *enter* a file name in the **File Name** field of the dialog before *clicking* the **Save** button. To open a drawing which has been saved *click* **Open** . . . in the **File** drop-down menu, and in the **Select File** dialog which appears select a file name form the file list.

There are differences between saving a 2D and a 3D drawing, in that when 3D model drawing is shaded by using a shading mode from the **3D Orbit/Shading Modes** sub-menu, the shading is saved with the drawing.

Exercises

1. A rendering of an assembled lathe tool holder is shown in Fig. 16.20. The rendering includes different materials for each part of the assembly.

    Working to the dimensions given in the parts orthographic drawing (Fig. 16.21), construct a 3D model drawing of the assembled lathe tool holder on several layers of different colours. Add lighting and materials and render the model in an isometric view. Shade with **3D Orbits/Hidden** and print or plot a **SW Isometric** view of the model drawing.

2. Working to the sizes given in the two-view orthographic projection in Fig. 16.22, construct a 3D model drawing of the drip tray from an engine. Add lighting and the material **OLIVE METAL**, place the model in an isometric view and render (Fig. 16.23).

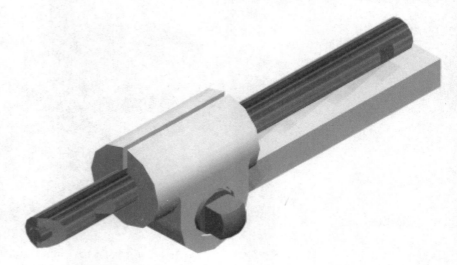

Fig. 16.20 Exercise 1 – a
rendering

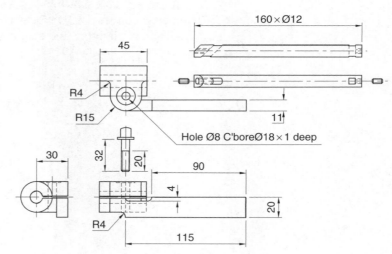

Fig. 16.21 Exercise 1 – parts
drawing

160 × Ø12

45

R4

R15

Hole Ø8 C'boreØ18 × 1 deep

11

30

32

20

90

4

20

115

R4

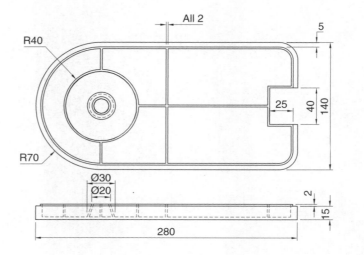

Fig. 16.22 Exercise 2 –
orthographic projection

All 2

R40

5

25

40

140

R70

Ø30

Ø20

2

15

280

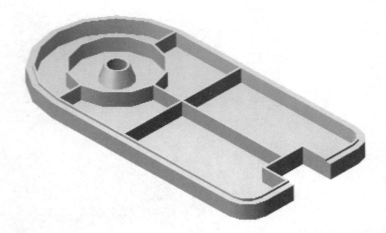

Fig. 16.23 Exercise 2

3. A three-view drawing of a hanging spindle bearing in third angle orthographic projection is shown in Fig. 16.24. Working to the dimensions in
the drawing, construct a 3D model drawing of the bearing.

Add suitable lighting and a suitable material, place in an isometric
view and render the model.

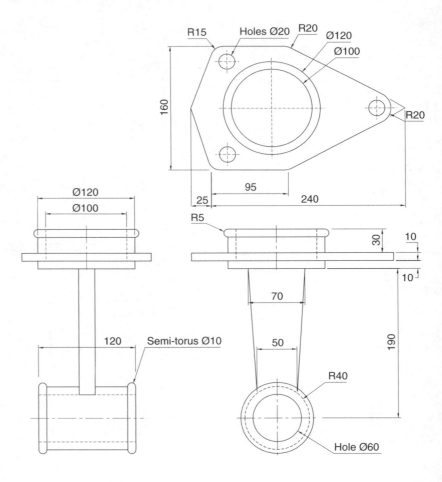

Fig. 16.24 Exercise 3

CHAPTER 17

Three-dimensional space

Aim of this chapter

To show in examples the methods of manipulating 3D models in 3D space using tools from the **UCS** toolbars.

3D space

So far in this book, when constructing 3D model drawings, they have been constructed on the AutoCAD 2005 coordinate system which is based upon three planes: the **XY Plane** – the screen of the computer, the **XZ Plane** at right angles to the **XY Plane** and as if coming towards the operator of the computer and a third plane (**YZ**) at right angles to both the other two planes (Fig. 17.1).

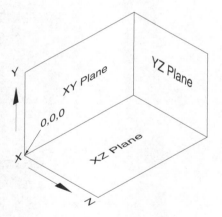

Fig. 17.1 The 3D space planes

In earlier chapters in order to view 3D objects which have been constructed on these three planes or at other angles, we have used presets from the **3D Views** sub-menu of the **Views** drop-down menu and have indicated other methods of rotating the model in 3D Space and placing the model in other viewing positions using the **Vpoint Presets** dialog.

Note

The **XY** plane is the basic **UCS** plane, which in terms of the ucs is known as the ***WORLD*** plane.

The User Coordinate System (UCS)

The **UCS** allows the operator to place the AutoCAD coordinate system in any position in 3D space using a variety of **UCS** tools (commands). Features of the **UCS** can be called either by *entering* **ucs** at the command line, or by selection from the **Tools** drop-down menu (Fig. 17.2), or from the two **UCS** toolbars – **UCS** and **UCS II** (Fig. 17.3).

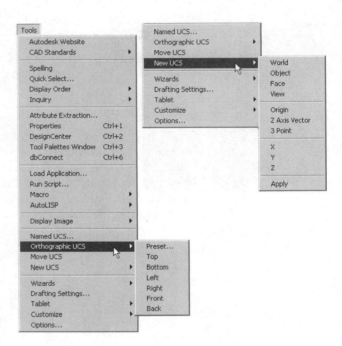

Fig. 17.2 The two sub-menus from the **Tools** drop-down menu

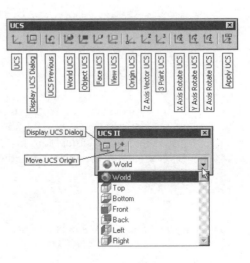

Fig. 17.3 The tools from the two **UCS** toolbars

If **ucs** is *entered* at the command line, it shows:

> **Command:** *enter* **ucs** *right-click*
> **Current ucs name: *WORLD***
> **Enter an option [New/Move/orthoGraphic/Prev/Restore/Save/Del/**
> **Apply/?/World] <World>:** *enter* **n** (New) *right-click*
> **Specify origin of UCS or [ZAxis/3point/OBject/Face/View/X/Y/Z]**
> **<0,0,0>:**

And from these prompt lines a selection can be made.

The two sub-menus – **Orthographic UCS** and **New UCS** – from the **Tools** drop-down menu provide a similar set of tools.

The tools may also be called from the **UCS** or **UCS II** toolbars.

The variable UCSFOLLOW

For the UCS to operate, the variable **UCSFOLLOW** must first be set on as follows:

> **Command:** *enter* **ucsfollow** *right-click*
> **Enter new value for UCSFOLLOW <0>:** *enter* **1** *right-click*
> **Command:**

The UCS icon

The **UCS** icon which indicates the direction of the three coordinate axes **X**, **Y** and **Z** is by default shown at the bottom left-hand corner of the AutoCAD drawing area as arrows pointing in the directions of the axes. When working in 2D, only the **X** and **Y** axes are shown, but when the drawing area is in a 3D view all three coordinate arrows are shown, except when the model is in the **XY** plane. The icon can be turned off as follows:

> **Command:** *enter* **ucsicon** *right-click*
> **Enter an option [ON/OFF/Noorigin/ORigin/Properties] <ON>:**

To turn the icon off, *enter* **off** in response to the prompt line and the icon disappears from the screen.

The appearance of the icon can be changed by *entering* **p** (Properties) in response to the prompt line. The **UCS Icon** dialog appears in which changes can be made to the shape, line width and colour of the icon if wished.

Types of UCS icon

The shape of the icon can be varied partly when changes are made in the **UCS Icon** dialog but also according to whether the AutoCAD drawing area is in 2D, 3D or Paper Space. A different form of 3D icon will appear when using the **3D Orbit** tool (Fig. 17.4).

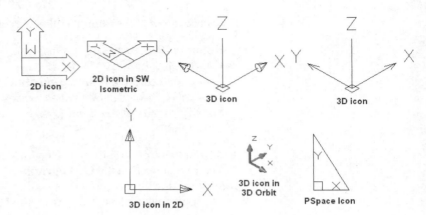

Fig. 17.4 Types of **UCS** icon

Examples of changing planes using the UCS

First example changing UCS planes (Fig. 17.6)

1. Set **UCSFOLLOW** to **1** (ON).
2. Place the screen in **3D Views/Front** and **Zoom** to **1**.
3. Construct the pline outline in Fig. 17.5 and extrude to a height of **120**.

Fig. 17.5 First example –
Changing UCS planes – pline
for extrusion

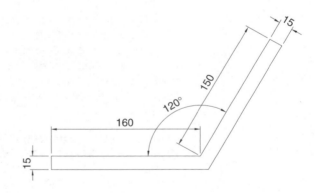

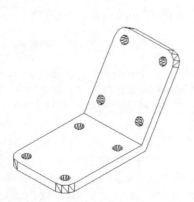

Fig. 17.6 First example –
Changing UCS planes

4. Place in the **SW Isometric** view and **Zoom** to **1**.
5. With the **Fillet** tool, fillet corners to a radius of **20**.
6. At the command line:

> **Command:** *enter* **ucs** *right-click*
> **Current ucs name: *WORLD***
> **Enter an option [New/Move/orthoGraphic/Prev/Restore/Save/Del/**
>   **Apply/?/World] <World>:** *enter* **n** (New) *right-click*
> **Specify origin of new UCS or [ZAxis/3point/OBject/Face/**
>   **View/X/Y/Z/]:** *enter* **f** (Face) *right-click*

Select face of solid object: *pick* the sloping face – its outline high-
   lights
Enter an option [Next/Xflip/Yflip] <accept>: *right-click*
Command:

And the 3D model changes its plane so that the sloping face is now
on the new UCS plane. **Zoom** to **1**.

7. On this new UCS, construct four cylinders of radius 7.5 and height −15
   (note the minus) and subtract them from the face.
8. *Enter* **ucs** at the command line again and *right-click* to place the
   model in the ***WORLD* UCS** – that being the plane in which the
   construction commenced.
9. Place four cylinders of the same radius and height into position in the
   base of the model and subtract them from the model.
10. Place the 3D model in the **SW Isometric** view and call **Hide**.

Second example – UCS (Fig. 17.9)

The 3D model for this example is a steam venting valve from a machine
of which a two-view third angle projection is shown (Fig. 17.7).

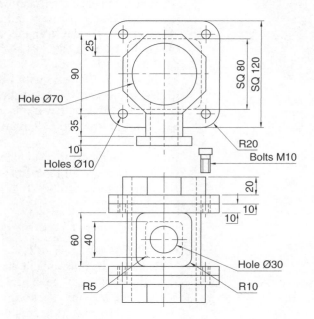

Fig. 17.7 Second example **UCS** –
the orthographic projection of a
steam venting valve

1. Make sure that **UCSFOLLOW** is set to **1**.
2. The **UCS** plane is the ***WORLD*** plane. Construct the **120** square
   plate at the base of the central portion of the valve. Construct five
   cylinders for the holes in the plate. Subtract the five cylinders from
   the base plate.
3. Construct the central part of the valve – a filleted **80** square extrusion
   with a central hole.

Fig. 17.8 Second example
UCS – construction up to item
11 + rendering

Fig. 17.9 Second example
UCS – items 12 and
13 + rendering

Fig. 17.10 Second example **UCS** –
pline for the bolt

4. Place the models in the **UCS orthoGonal/Front** plane.
5. With the **Move** tool, move the central portion vertically up by **10**.
6. With the **Copy Object** tool, copy the base up to the top of the central portion.
7. With the **Union** tool form a single 3D model of the three parts.
8. Make the layer **Construction** current.
9. Place the model in the **UCS *WORLD*** plane. Construct the separate top part of the valve – a plate forming a union with a hexagonal plate and with holes matching those of the other parts.
10. Place the scene so far in the **UCS orthoGonal/Front** plane and move the parts of the top into their correct positions relative to each other and with **Union** and **Subtract** tools, complete the part. This will be made easier if the layer **0** is turned off.
11. Turn layer **0** on and move the top into its correct position relative to the main part of the valve. Then with the **Mirror** tool, mirror the top to produce the bottom of the assembly (Fig. 17.8).
12. While in the ***FRONT* UCS** construct the three parts of a 3D model of the extrusion to the main body.
13. In the **UCS *WORLD*** move the parts into their correct positions relative to each other and with **Union** form a union of the two filleted rectangular extrusions and the main body. Then with **Subtract**, subtract the cylinder from the whole (Fig. 17.9).
14. In the **UCS *FRONT*** plane, construct one of the bolts as shown in Fig. 17.10, forming a solid of revolution from a pline. Then add a head to the bolt and with **Union** add it to the screw.
15. With the **Copy Object** tool, copy the bolt 7 times to give 8 bolts. With **Move**, and working in the **UCS *WORLD*** and ***FRONT*** planes, move the bolts into their correct positions relative to the 3D model.
16. Save the model to a suitable file name.
17. Finally move all the parts away from each other to form an exploded view of the assembly (Fig. 17.11).

Third example – UCS (Fig. 17.15)

1. Set **UCSFOLLOW** to **1**.
2. Place the drawing area in the **UCS *FRONT*** view.
3. Construct the outline in Fig. 17.12 and extrude to a height of **120**.
4. Either *click* the **3 Point UCS** tool icon in the **UCS** toolbar (Fig. 17.13) or at the command line (Fig. 17.14 shows the three UCS points):

   Command: *enter* **ucs** *right-click*
   Current ucs name: *RIGHT*
   Enter an option [prompts]: *enter* **n** (New) *right-click*
   Specify origin of UCS or [prompts]: *enter* **3** (3point) *right-click*
   Specify new origin point: *pick*
   Specify point on positive portion of X-axis: *pick*

Fig. 17.11 Second example **UCS**

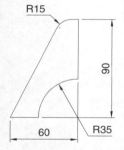

Fig. 17.12 Third example **UCS** – outline for 3D model

Fig. 17.13 The **3 Point UCS** tool icon in the **UCS** toolbar

Fig. 17.14 Third example **UCS** – the three UCS points

Specify point on positive-Y portion of UCS XY plane: *enter* **.xy** *right-click*
of *pick* (**need Z**): *enter* **−1** (note minus 1) *right-click*
Regenerating model.
Command:

And the model regenerates in this new 3point plane.

5. On the face of the model construct a rectangle **80 × 50** central to the face of the front of the model, fillet its corners to a radius of **10** and extrude to a height of **10**.
6. Place the model in the **SW Isometric** view and fillet the back edge of the second extrusion to a radius of **10**.
7. Subtract the second extrusion from the first.
8. Add lights and the material **WOOD MED_ASH** and render the model (Fig. 17.15).

Fourth example – UCS (Fig. 17.17)

1. With the last example still on screen, place the model in the **UCS** ***WORLD*** view.
2. *Click* the **Z Axis Vector UCS** tool icon in the **UCS** toolbar (Fig. 17.16). The command line shows:

Command:_ucs
Current ucs name: *WORLD*
Enter an option [prompts] <World>:_zaxis
Specify a new origin point: *enter* **40,60** *right-click*
Specify point on positive portion of Z-axis <40,60,0>: *enter* **.xy** *right-click*
of *enter* **170,220** *right-click* (**need Z**): *enter* **1** *right-click*
Regenerating model.
Command:

3. Render the model in its new **UCS** plane (Fig. 17.17).

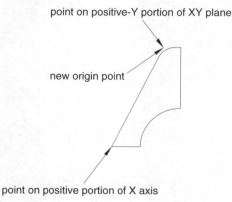

point on positive-Y portion of XY plane

new origin point

point on positive portion of X axis

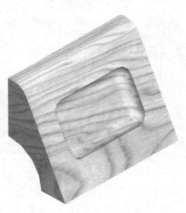

Fig. 17.15 Third example **UCS**

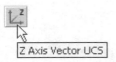

Fig. 17.16 The **Z Axis Vector UCS** icon in the **UCS** toolbar

Fig. 17.18 The **UCS** icon in the **UCS** toolbar

Fig. 17.19 The **Display UCS Dialog** icon in the **UCS II** toolbar

Saving UCS views

If a number of different **UCS** planes are used in connection with the construction of a 3D model, each can be saved to a different name and recalled when required. To save the UCS plane in which a 3D model drawing is being constructed, either *click* the **UCS** tool icon in the **UCS** toolbar (Fig. 17.18) or *enter* **ucs** at the command line:

Command:_ucs
Current ucs name: NW Isometric
Enter an option [prompts]: *enter* **s** (Save) *right-click*
Enter name to save current UCS: *enter* **SW Isometric** *right-click*
Regenerating model.
Command:

Fig. 17.17 Fourth example **UCS**

Now *click* the **Display UCS Dialog** tool icon in the **UCS II** toolbar and the **UCS** dialog appears (Figs. 17.19 and 17.20) showing the names of the views saved in the current drawing.

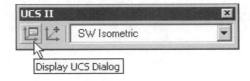

Constructing 2D objects in 3D space

In previous pages of this book there have been examples of 2D objects constructed with the **Polyline**, **Line**, **Circle** and other 2D tools to form the outlines for extrusions and solids of revolution. These outlines have been drawn on planes set either from the **3D Views** sub-menu of the **View** drop-down menu, or in **UCS** planes such as the **UCS *RIGHT***, ***FRONT*** and ***LEFT*** planes.

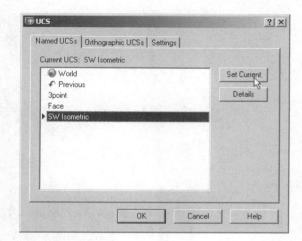

Fig. 17.20 The **UCS** dialog

First example – 2D outlines in 3D space (Fig. 17.23)

1. Construct a **3point UCS** to the following points:

   origin point: 80,90
   X-axis point: 290,150
   positive-Y point: .xy of 80,90
   (need Z): *enter* **1**.

2. On this **3point UCS** construct a 2D drawing of the plate to the dimensions given in Fig. 17.21, using the **Polyline**, **Ellipse** and **Circle** tools.

Fig. 17.21 First example – **2D outlines in 3D space**

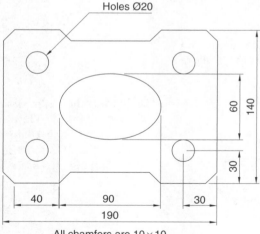

Holes Ø20

All chamfers are 10×10

Fig. 17.22 First example – **2D outlines in 3D space**. The outline in a **SW Isometric** view

3. Save the **UCS** plane in the **UCS** dialog to the name **3point**.
4. Place the drawing area in the **SW Isometric** view (Fig. 17.22).
5. With the **Region** tool form regions of the six parts of the drawing and with the **Subtract** tool, subtract the circles and ellipse from the main outline.
6. Extrude the region to a height of **10** (Fig. 17.23).

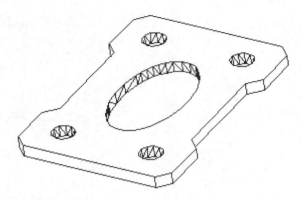

Fig. 17.23 First example – **2D
outlines in 3D space**

Second example – 2D outlines in 3D space (Fig. 17.26)

1. Place the drawing area in the **UCS *FRONT*** view, **Zoom** to **1** and construct the outline in Fig. 17.24.

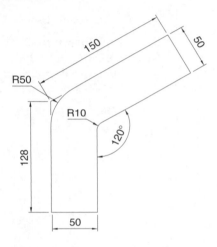

Fig. 17.24 Second example – **2D
outlines in 3D space**. Outline
to be extruded

Face UCS

Fig. 17.25 The **Face UCS** icon
from the **UCS** toolbar

2. Extrude the outline to a height of **150**.
3. Place in the **3D Views/SW Isometric** view and **Zoom** to **1**.
4. *Click* the **Face UCS** tool icon in the **UCS** toolbar (Fig. 17.25) and place the 3D model in the ucs plane shown in Fig. 17.26, selecting the sloping face of the extrusion for the plane and again **Zoom** to **1**.

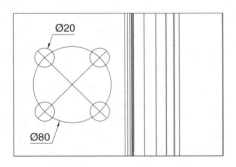

Fig. 17.26 Second example – **2D
outlines in 3D space** – the
circles in the new UCS face

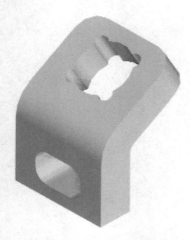

Fig. 17.27 Second example – **2D outlines in 3D space**

5. With the **Circle** tool draw five circles as shown in Fig. 17.26.
6. Form a region from the five circles and with **Union** form a union of the regions.
7. Extrude the region to a height of **−60** (Note the minus) – higher than the width of the sloping part of the 3D model.
8. Place the model in the **SW Isometric** view and subtract the extruded region from the model.
9. With the **Fillet** tool, fillet the upper corners of the slope of the main extrusion to a radius of **30**.
10. Place the model in another **UCS FACE** plane and construct a filleted pline of sides **80** and **50** and filleted to a radius of **20**. Extrude to a height of **−60** and subtract the extrusion from the 3D model.
11. Place in the **SW Isometric** view and render (Fig. 17.27).

Revision notes

1. The **UCS** (User Coordinate System) tools can be called from the two toolbars **UCS** and **UCS II** or from sub-menus from the **Tools** drop-down menu.
2. The variable **UCSFOLLOW** must first be set on (to **1**) before operations of the UCS can be brought into action.
3. There are several types of **UCS** icons – 2D (different types), 3D (different types), 3D Orbit, Pspace.
4. The position of the plane in 3D space on which a drawing is being constructed can be varied using tools from the **UCS**.
5. The different planes on which drawings are constructed in 3D space can be saved in the **UCS** dialog.

Fig. 17.28 Exercise 1 – a rendering

Exercises

1. The two-view projection shows an angle bracket in which two pins are placed in holes in each of the arms of the bracket as shown in Fig. 17.29. Construct a 3D model of the bracket and its pins. Add lighting to the scene and materials to the parts of the model and render (Fig. 17.28).
2. The two-view projection (Fig. 17.31) shows a stand consisting of two hexagonal prisms. Circular holes have been cut right through each face of the smaller hexagonal prism and rectangular holes with rounded ends have been cut right through the faces of the larger.
   Construct a 3D model of the stand. When completed add suitable lighting to the scene. Then add a material to the model and render (Fig. 17.30).
3. The two-view projection (Fig. 17.32) shows a ducting pipe. Construct a 3D model drawing of the pipe. Place in a **SW Isometric** view, add lighting to the scene and a material to the model and render.
4. A point marking device is shown in the two-view projection in Fig. 17.33. The device is composed of three parts – a base, an arm and a pin.
   Construct a 3D model of the assembled device and add appropriate materials to each part. Then add lighting to the scene and render in a **SW Isometric** view (Fig. 17.34).

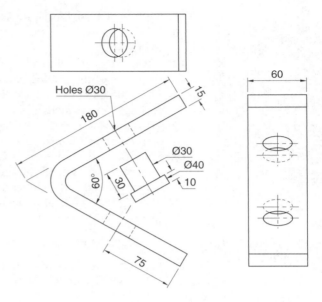

Fig. 17.29 Exercise 1 – details of
the shapes and sizes

Fig. 17.30 Exercise 2 – a rendering

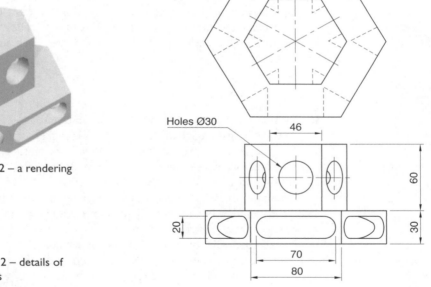

Fig. 17.31 Exercise 2 – details of
the shapes and sizes

5. Fig. 17.35 shows the rendering of a 3D model drawing of the connect-
    ing device shown in the orthographic projection in Fig. 17.36. Con-
    struct the 3D model drawing of the device and add a suitable lighting
    to the scene. Then place in a **SE Isometric** view, add a material to the
    model and render.

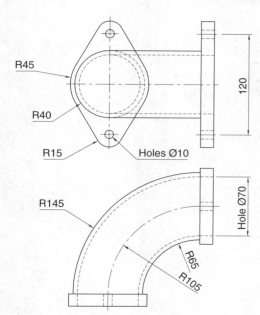

Fig. 17.32 Exercise 3 – details of
the shapes and sizes

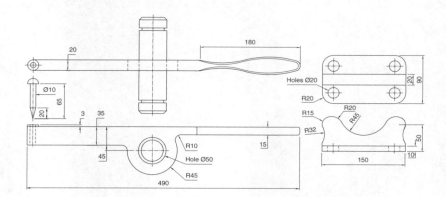

Fig. 17.33 Exercise 4 – details of
the shapes and sizes

Fig. 17.34 Exercise 4 – a rendering

Fig. 17.35 Exercise 5 – a rendering

Fig. 17.36 Exercise 5 – two-view drawing

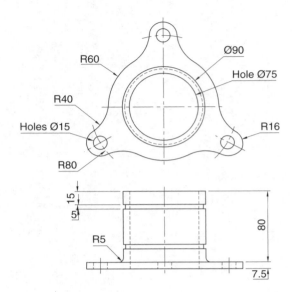

6. A fork connector and its rod are shown in a two-view projection (Fig. 17.37). Construct a 3D model drawing of the connector with its rod in position. Then add lighting to the scene, place in an **Isometric** viewing position, add materials to the model and render.

Fig. 17.37 Exercise 6

Fig. 17.38 Exercise 7 – a rendering

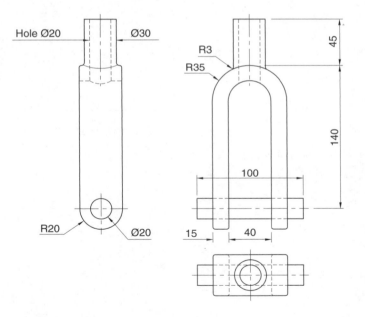

7. An orthographic projection of the parts of a lathe steady are given in Fig. 17.39. From the dimensions shown in the drawing, construct an assembled 3D model of the lathe steady (Fig. 17.40). When the 3D model has been completed, add suitable lighting and materials and render the model as shown in Fig. 17.38.

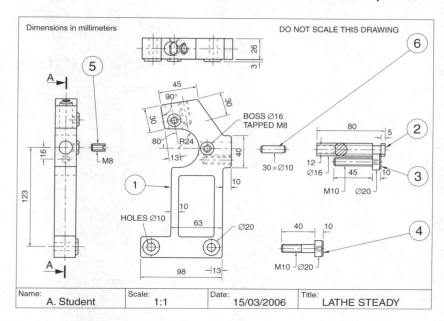

Fig. 17.39 Exercise 7 – details

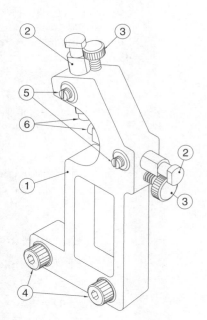

Fig. 17.40 Exercise 7 – an
isometric drawing

CHAPTER 18

3D surface models

Aims of this chapter

1. To introduce the idea of 3D surfaces.
2. To compare 3D surface models with 3D solid drawing models.
3. To give examples of 3D surface models.

3D surface meshes

3D surface models will be introduced in this chapter. The 3D models described in earlier chapters have all be constructed using the tools from the **Solids** toolbar. 3D surface models are constructed using tools from the **Surfaces** toolbar (see page 165). 3D surface model drawings are constructed from 3D surface meshes. In some instances it appears as if 3D solid models are the same as 3D surface models, but in fact there are distinct differences.

1. The surface meshes of 3D solid models are controlled by the settings of the variable **ISOLINES**.
2. The surface meshes of 3D surface models are controlled either by the number of segments or by the two variables **SURFTAB1** and/or **SURFTAB2**.
3. The Boolean operators **Union**, **Subtract** and **Intersect** can be used to join, subtract or intersect 3D model objects, but have no action when used with 3D surface objects.

Comparisons between Solids and Surfaces tools

First example – comparing 3D solid and 3D surface models (Fig. 18.3)

The three 3D cubes shown in the upper drawing of Fig. 18.3 have been constructed as follows:

Left-hand cube: using the **Line** tool.
Central cube: using the **Box** tool from the **Solids** toolbar (Fig. 18.1).
Right-hand cube: using the **Box** tool from the **Surfaces** toolbar (Fig. 18.2).

The lower three drawings of Fig. 18.3 show the results of calling the **Hide** tool on each of the three upper drawings.

Fig. 18.1 The **Box** tool icon from the **Solids** toolbar

Fig. 18.2 The **Box** tool icon from the **Surfaces** toolbar

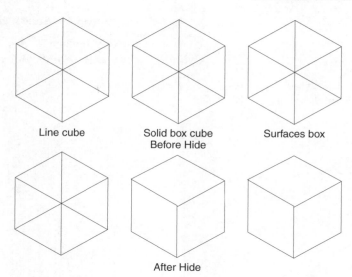

Line cube

Solid box cube
Before Hide

Surfaces box

Fig. 18.3 First example – comparing 3D solid and 3D surface models

After Hide

Second example – Surfaces tool – Cone (Fig. 18.5)

The three cylinders were all constructed using the **Cone** tool from the **Surfaces** toolbar (Fig. 18.4).

Fig. 18.4 The **Cone** tool icon from the **Surfaces** toolbar

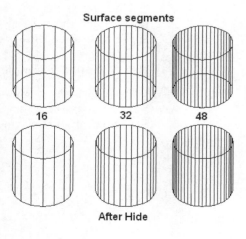

Surface segments

16   32   48

After Hide

Fig. 18.5 Second example – **Surfaces** tool – **Cone**

When using the **Cone** tool, the last prompt line is:

Enter number of segments for surface of cone: *enter* a figure *right-click*

The upper three cones (top and bottom of cone are of same radius) show the number of segments entered in response to this prompt line.

Note that when **Hide** is called, the upper surface of the cylinders do not have surface meshes and so are open.

Fig. 18.6 The **Cylinder** tool icon
from the **Solids** toolbar

Third example – Solids tool – Cylinder (Fig. 18.7)

The upper three cylinders in this example were constructed using the
Cylinder tool from the **Solids** toolbar (Fig. 18.6).

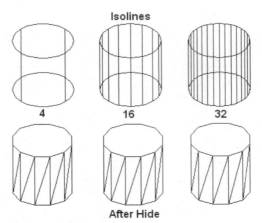

Fig. 18.7 Third example – **Solids**
tool – **Cylinder**

The lower three drawings show the results of calling **Hide** to the three
cylinders in the upper drawings.

Note that the upper surface of the three cylinders hide the rest of the
lines behind the top and that all three cylinders assume the same mesh
appearance when **Hide** is called.

Fourth example – Surfaces tool – Edgesurf (Fig. 18.8)

When constructing 3D surface models using tools such as the **Edge
Surface** tool from the **Surfaces** toolbar, both the **Surftab1** and the
Surftab2 variables may need resetting.

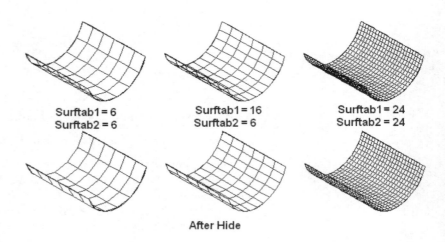

Fig. 18.8 Fourth example –
Surfaces tool – **Edgesurf**

The upper three drawings of Fig. 18.8 show 3D edgesurf models with a note under each stating the **Surftab** settings; the lower three models show the three surface models after **Hide** has been called.

The Surface tools

The **Surfaces** toolbar showing the names of the tools represented by the tool icons was shown on page 165. The following examples show the use of many of the tools in this toolbar.

Fig. 18.9 The **Cone** tool icon from the **Surfaces** toolbar

First example – 3D Surfaces model (Fig. 18.12)

1. *Click* on the **Cone** tool icon in the **Surfaces** toolbar (Fig. 18.9). The command line shows:

   Command:_ai_cone
   Specify center point for base of cone: *enter* **190,170** *right-click*
   Specify radius for base of cone: *enter* **30** *right-click*
   Specify radius for top of cone: *enter* **30** *right-click*
   Specify height of cone: *enter* **80** *right-click*
   Enter number of segments for surface of cone <16>: *right-click*
   Command:

Fig. 18.10 The **Box** tool icon from the **Surfaces** toolbar

2. *Click* on the **Box** tool icon in the **Surfaces** toolbar (Fig. 18.10). The command line shows:

   Command:_ai_box
   Specify corner point of box: *enter* **160,140,80** *right-click*
   Specify length of box: *enter* **60** *right-click*
   Specify width of box: *enter* **60** *right-click*
   Specify height of box: *enter* **30** *right-click*
   Specify rotation angle of box about the Z axis: *enter* **0** *right-click*
   Command:

Fig. 18.11 The **Dome** tool icon from the **Surfaces** toolbar

3. *Click* on the **Dome** tool icon in the **Surfaces** toolbar (Fig. 18.11). The command line shows:

   Command:_ai_dome
   Specify center point of dome: *enter* **190,170,110** *right-click*
   Specify radius of dome: *enter* **30** *right-click*
   Enter number of longitudinal segments for surface of dome <16>: *right-click*
   Enter number of latitudinal segments for surface of dome <8>: *enter* **16** *right-click*
   Command:

4. Place the model in the **SW Isometric** view.
5. Call **Hide**.

The resulting 3D surfaces solid is shown in Fig. 18.12.

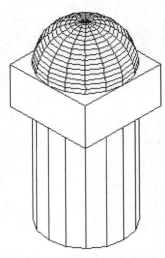

Fig. 18.12 First example – **3D Surfaces model**

Torus

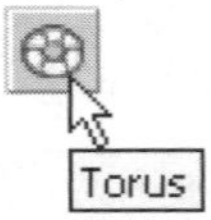

Fig. 18.13 The **Torus** tool icon from the **Surfaces** toolbar

Note

Try forming a union of the three 3D surface objects by *clicking* the **Union** tool from the **Solids Editing** toolbar. The command line will show:

Command:_union
Select objects: *pick* the cylinder **1 found**
Select objects: *pick* the box **1 found, 2 total**
Select objects: *pick* the dome **1 found, 3 total**
Select objects: *right-click*
At least 2 solids or coplanar regions must be selected.
Command:

It will be seen that the three surface models will not form a union. Test by moving any one of the three surface models.

Second example – Surfaces model (Fig. 18.14)

1. Construct the same cylinder (using the **Cone** tool from **Surfaces**) as for the first example.
2. *Click* on the **Torus** tool in the **Surfaces** toolbar (Fig. 18.13). The command line shows:

Command:_ai_torus
Specify center point of torus: *enter* **190,170,25** *right-click*
Specify radius of torus: *enter* **35** *right-click*
Specify radius of tube: *enter* **5** *right-click*
Enter number of segments around the tube circumference <16>: *right-click*
Enter number of segments around torus circumference <16>: *right-click*
Command:

3. Construct a similar torus at a height of **55** above the **XY** plane.
4. Add a **Dome** of radius **30** at a height above the **XY** plane of **80**.
5. Place the model in the **SW Isometric** view. **Zoom** to **1**.
6. Call **Hide**.

Third example – Surfaces model (Fig. 18.17)

1. Construct the plines as shown in Fig. 18.15.
2. Set **SURFTAB1** to **24**.
3. Set **SURFTAB2** to **6**.
4. *Click* the **Revolved Surface** tool icon in the **Surfaces** toolbar (Fig. 18.16). The command line shows:

Command:_revsurf
Current wire frame density: SURFTAB1=24 SURFTAB2=6
Select object to revolve: *pick* the curved pline
Select object that defines the axis of revolution: *pick*
Specify start angle <0>: *right-click*
Specify included angle <360>: *right-click*
Command:

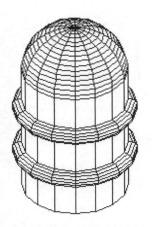

Fig. 18.14 Second example – **Surfaces model**

Fig. 18.15 Third example –
Surfaces model – pline outline

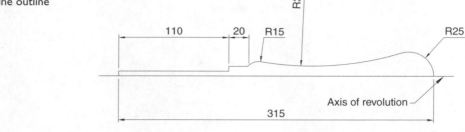

Fig. 18.16 The **Revolved Surface** tool icon from the **Surfaces** toolbar

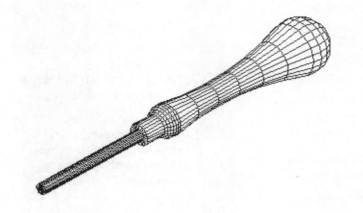

Fig. 18.17 Third example –
Surfaces model

5. Erase the axis of revolution pline.
6. Place in a **SW Isometric** view, **Zoom** to **1** and call **Hide**.

Third example – Surfaces model (Fig. 18.19)

1. Set **Surftab1** to **2**.
2. In the **UCS *WORLD*** construct a hexagon of edge length **35**.
3. In the **UCS *FRONT*** and in the centre of the hexagon construct a pline of height **100**.
4. Place the drawing area in the **SW Isometric** view.
5. *Click* on the **Tabulated Surface** tool in the **Surfaces** toolbar (Fig. 18.18). The command line shows:

Fig. 18.18 The **Tabulated Surface** tool icon from the **Surfaces** toolbar

   Command:_tabsurf
   Current wire frame density: SURFTAB1=6
   Select objects for path curve: *pick* the hexagon
   Select object for direction vector: *pick* the pline
   Command:

6. Call **Hide**.

Fourth example – Surfaces model (Fig. 18.22)

1. In the **UCS *FRONT*** construct the pline as shown in Fig. 18.20.
2. In the **UCS *WORLD***, **Zoom** to **1** and copy the pline to a vertical distance of **120**.

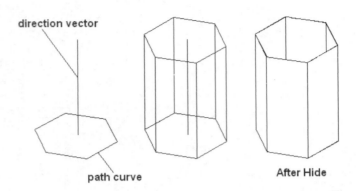

direction vector

path curve

After Hide

Fig. 18.19 Third example –
Surfaces model

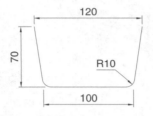

Fig. 18.20 Fourth example –
Surfaces model – pline outline

Ruled Surface

Fig. 18.21 The **Ruled Surface**
tool icon from the **Surfaces**
toolbar

3. Place in the **SW Isometric** view and **Zoom** to **1**.
4. Set **SURFTAB1** to **32**.
5. *Click* on the **Ruled Surface** tool in the **Surfaces** toolbar (Fig. 18.21).
The command line shows:

Command:_rulesurf
Current wire frame density: SURFTAB1=32
Select first defining curve: *pick* one of the plines
Select second defining curve: *pick* the other pline
Command:

6. Call **Hide**.

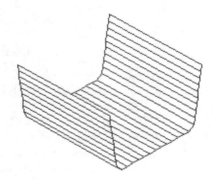

Fig. 18.22 Fourth example –
Surfaces model

Fifth example – Surfaces model (Fig. 18.26)

1. Place the drawing area in the **UCS *RIGHT***. **Zoom** to **1**.
2. Construct the polyline to the sizes and shape as shown in Fig. 18.23.

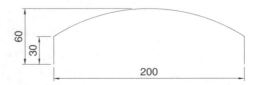

Fig. 18.23 Fifth example –
Surfaces model – pline outline

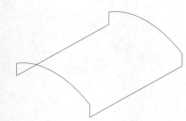

Fig. 18.24 Fifth example –
Surfaces model – adding lines
joining the outlines

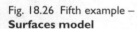

Fig. 18.25 The **Edge Surface**
tool icon from the **Surfaces**
toolbar

3. Place the drawing area in the **UCS *WORLD***. **Zoom** to **1**.
4. Copy the pline to the right by **250**.
5. Place the drawing in the **SW Isometric** view and **Zoom** to **1**.
6. With the **Line** tool, draw lines between the ends of the two plines using the **endpoint** osnap (Fig. 18.24). Note that if polylines are drawn they will not be accurate at this stage.
7. Set **SURFTAB1** to **32** and **SURFTAB2** to **64**.
8. *Click* on the **Edge Surface** tool icon in the **Surfaces** toolbar (Fig. 18.25). The command line shows:

Command:_edgesurf
Current wire frame density: SURFTAB1=32 SURFTAB2=64
Select object 1 for surface edge: *pick* one of the lines (or plines)
Select object 1 for surface edge: *pick* the next adjacent line (or pline)
Select object 1 for surface edge: *pick* the next adjacent line (or pline)
Select object 1 for surface edge: *pick* the last line (or pline)
Command:

9. Call **Hide**.

The result is shown in Fig. 18.26

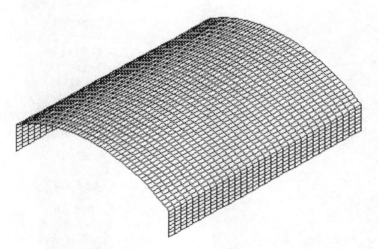

Fig. 18.26 Fifth example –
Surfaces model

The action of Pedit on a 3D surface (Fig. 18.27)

Open the drawing file saved for the last 3D drawing model (the edgesurf model) and copy it twice. Then call the **Polyline Edit** tool. The command line shows:

Command:_pedit
Select polyline or [Multiple]: *pick* the first copy
Enter an option [Edit vertex/Smooth surface/Desmooth/Mclose/
  Nclose/Undo]: *enter* **s** (Smooth surface) *right-click*
Enter an option [Edit vertex/Smooth surface/Desmooth/Mopen/
  Nclose/Undo]: *right-click*
Command:

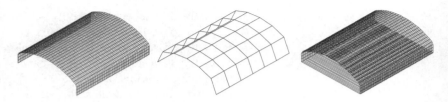

Fig. 18.27 The action of **Pedit** on a 3D surface model

And the 3D surface changes to the view given in the middle drawing in Fig. 18.27.

If, in answer to the first prompt line, an **n** (for Nclose) is *entered*, the result will be as shown in the right-hand drawing.

These results can be changed back by using either the **d** (for Desmooth) for the central drawing or **m** (for Mopen) for the right-hand drawing.

Fig. 18.28 The **2D Solid** tool icon from the **Surfaces** toolbar

Notes on the 3D Surface tools

1. The first tool icon in the toolbar is the **2D Solid** tool (Fig. 18.28). This tool is not a surfaces tool as such. Outlines constructed with the aid of this tool will be filled with black.
2. The second tool is **3D Face**. This tool has been used in examples and assignments in Chapter 13.
3. The **3D Mesh** tool (Fig. 18.29) is mainly used by those developing tools for use with AutoCAD and will not be dealt with here.
4. The **Edge** tool (Fig. 18.30) can be used if necessary for removing edges from 3D faces.

Fig. 18.29 The **3D Mesh** tool icon from the **Surfaces** toolbar

Fig. 18.30 The **Edge** tool icon from the **Surfaces** toolbar

Rendering of 3D Surface models

As with 3D solid model drawings, 3D surface model drawings can be rendered if thought necessary. By adding lights and materials to the surfaces, surface models can be rendered as effectively as 3D solid models.

Examples of surface renderings involving lights and a material showing the action of the **Polyline Edit** tool on copies of the fifth example are shown in Fig. 18.31.

Fig. 18.31 Rendering of 3D surface models

Revision notes

1. The density of 3D surface meshes of surface models are controlled by the two set variables **SURFTAB1** and **SURFTAB2**.
2. The **Solids** tools **Union**, **Subtract** and **Intersect** (the Boolean operators) cannot be used with surface models.
3. When constructing a surface 3D cylinder, the **Cone** tool from the **Surfaces** toolbar must be used.
4. Surfaces formed with the **Ruled Surface** tool require two outlines from which the surface can be obtained.
5. Surfaces formed with the **Edge Surface** tool require four edges from which to form the surface. The edges must meet each other at their ends for the surface to form.
6. 3D surface models can be rendered using the same methods as are used when rendering 3D solid models.

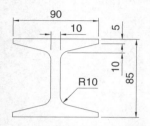

Fig. 18.32 Exercise 1 – the outline from which the surface model is formed

Exercises

1. In the **UCS *WORLD*** construct the polyline outline as shown in Fig. 18.32. With the **Tabulated Surface** tool create a girder of length **200** from the outline (left-hand drawing of Fig. 18.33). Using the **Ruled Surface** tool, construct a surface to cover the top end of the girder. Render the girder after adding suitable lights and the material **COPPER** (right-hand drawing of Fig. 18.33).

Fig. 18.33 Exercise 1 – the **Tabulated Surface** and the rendering

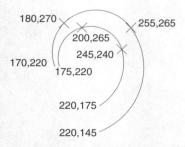

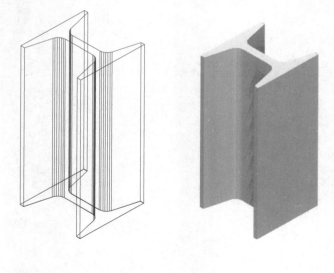

Fig. 18.34 Exercise 2 – the arcs on which the exercise is based

2. Using the **Polyline** tool construct arcs as shown in Fig. 18.34. Using the **UCS 3point** option construct semicircles joining both ends

Fig. 18.35 Exercise 2 –
semi-circles at end of arcs

of the arcs as shown in Fig. 18.35. With the **Edge Surface** tool and with the **SURFTAB** variables set to suitable sizes, construct a surface from the four arcs (Fig. 18.36). Using the **Mirror** tool, mirror the surface twice to obtain the hook shape as shown in the rendering in Fig. 18.37. Render the hook after adding lighting and the material **BRASS GIFMAP**.

3. Construct a surface of revolution from the polyline outline given in the left-hand drawing of Fig. 18.38. Add lighting and the material **BLUE GLASS** and after placing the surface model in a suitable isometric viewing position render the model – right-hand illustration (Fig. 18.38).

Fig. 18.36 Exercise 2 – a surface

Fig. 18.37 Exercise 2

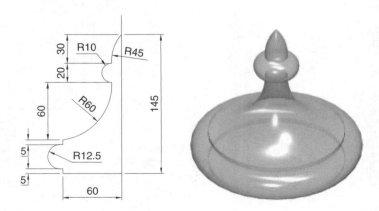

Fig. 18.38 Exercise 3

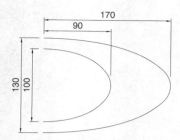

Fig. 18.39 Exercise 4 – the two semi-ellipses

4. In the **UCS *WORLD*** plane construct two semi-ellipses to the sizes as given in Fig. 18.39. Place in the **UCS *FRONT*** plane and move the smaller semi-ellipse vertically upwards by **200** units. Using the **Ruled Surface** tool, form a surface between the two semi-ellipses. Add suitable lighting and a suitable material and render the surface (Fig. 18.40).

5. Using the **Pyramid** and **Box** tools from the **Surfaces** toolbar, construct the 3D model in Fig. 18.41. Then place the model in the **SW Isometric** view and using the **Hidden** shading from the **3D Orbit** *right-click* menu, hide lines behind the front faces (Fig. 18.42).

Fig. 18.40 Exercise 4

Fig. 18.41 Exercise 5 – sizes of the parts of the model

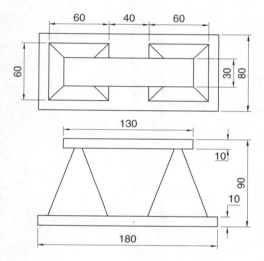

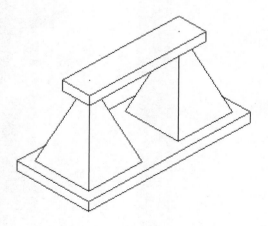

Fig. 18.42 Exercise 5

6. Working to the sizes given in the front view in Fig. 18.43, construct the 3D surface model shown in the rendering in Fig. 18.44.

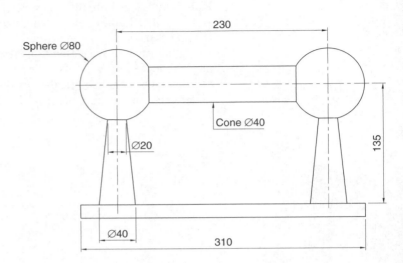

Fig. 18.43 Exercise 6 – front view

Fig. 18.44 Exercise 6

CHAPTER 19

Editing 3D solid models: More 3D models

Aims of this chapter

1. To introduce the use of tools from the **Solids Editing** toolbar.
2. To give examples of the use of the tools from the **Solids Editing** toolbar.
3. To show examples of a variety of 3D solid and 3D surface models.

The Solids Editing tools

The tools from the **Solids Editing** toolbar have already been illustrated on page 165. Examples of the results of using some of the tools from this toolbar are shown in this chapter. These tools are of value if the design of a 3D solid model requires to be changed (edited), although some have a value in constructing solids which cannot easily be constructed using other tools. The **Solids Editing** tools have no effect when used on 3D surface models.

First example – Extrude Faces tool (Fig. 19.3)

1. Set **ISOLINES** to 24.
2. In the **UCS *RIGHT*** plane construct a cylinder of radius **30** and height **30** (Fig. 19.1).

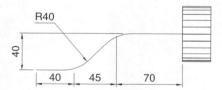

Fig. 19.1 First example –
Extrude Faces tool – first stages

3. In the **UCS *FRONT*** view construct the pline in Fig. 19.1. Mirror the pline to the other end of the cylinder.
4. In the **UCS *WORLD*** move the pline to lie central to the cylinder.
5. Place the screen in the **SW Isometric** view.
6. *Click* the **Extrude Faces** tool icon in the **Solids Editing** toolbar (Fig.19.2). The command line shows:

Command:_solidedit
Enter a solids editing option [Face/Edge/Body/Undo/eXit]:_face

Fig. 19.2 The **Extrude Faces**
tool icon from the **Solids
Editing** toolbar

259

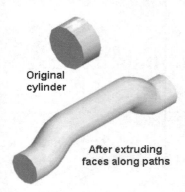

Original
cylinder

After extruding
faces along paths

Fig. 19.3 First example –
Extrude Faces tool

Enter a face editing option [Extrude/Move/Rotate/Offset/Taper/
Delete/Copy/coLor/Undo/eXit]:_extrude
Select faces or [Undo/Remove/ALL]: *pick* the left-hand face of the
cylinder **2 faces found**
Select faces or [Undo/Remove/ALL]: *pick* highlighted faces which
are not to be extruded **1 found. 1 removed**
Select faces or [Undo/Remove/ALL]: *right-click*
Specify height of extrusion or [Path]: *enter* **p** *right-click*
Select extrusion path: *pick* the pline
Solids editing automatic checking. SOLIDCHECK=0

7. Repeat the operation using the plane at the other end of the cylinder.
8. Render the resulting edited 3D model (Fig. 19.3).

Notes

1. Note the prompt line which includes the statement **SOLIDCHECK=0**.
If the variable **SOLIDCHECK** is set on (to **1**) the prompt lines change to
include the lines **SOLIDCHECK=1**, **Solid validation started** and
Solid validation completed.
2. When a face is *picked* other faces become highlighted, using the
Remove option of the prompt line **Select faces or [Undo/Remove/
ALL]** allows faces which are not to be extruded to be removed from
the operation of the tool.

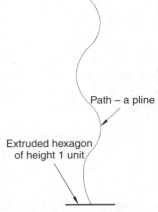

Path – a pline

Extruded hexagon
of height 1 unit

Fig. 19.4 Second example –
Extrude Faces tool – pline
for path

Second example – Extrude Faces tool

1. Construct a hexagonal extrusion just **1** unit high in the **UCS
*WORLD*** plane.
2. Change to the **UCS *FRONT*** plane and construct the curved pline in
Fig. 19.4.
3. Back in the **UCS *WORLD*** move the pline to lie central to the extrusion.
4. Place in the **SW Isometric** view and extrude the top face of the extrusion
along the path of the curved pline.
5. Add lighting and a material to the model and render (Fig. 19.5).

Note

This example shows that a face of a 3D solid model can be extruded
along any suitable path curve. If the polygon on which the extrusion
had been based had been turned into a region, no extrusion could have
taken place. The polygon had to be extruded to give a face to the 3D
solid.

Third example – Move Faces tool (Fig. 19.7)

1. Construct the 3D solid drawing shown in the left-hand drawing of
Fig. 19.7 from three boxes which have been united using the **Union**
tool.

Fig. 19.5 Second example –
Extrude Faces tool

Fig. 19.6 The **Move Faces** tool
icon from the **Solids Editing**
toolbar

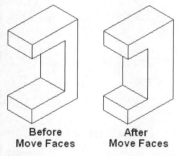

Before Move Faces   After Move Faces

Fig. 19.7 Third example – **Move Faces** tool

2. *Click* on the **Move Faces** tool in the **Solids Editing** toolbar (Fig. 19.6). The command line shows:

Command:_solidedit
[prompts]:_face
Enter a face editing option
[prompts]:_move
Select faces of [Undo/Remove/ALL]: *pick* the face to be moved **2 faces found**
Select faces of [Undo/Remove/ALL]: *enter* **r** (remove) *right-click*
Select faces of [Undo/Remove/ALL]: *pick* **2 faces found, 1 removed**
Select faces of [Undo/Remove/ALL]: *right-click*
Specify a base point or displacement: *pick*
Specify a second point of displacement: *pick*
[further prompts]:

And the *picked* face is moved – right-hand drawing of Fig. 19.7.

Fourth example – Offset Faces tool (Fig. 19.8)

1. Construct the 3D solid drawing shown in the left-hand drawing of Fig. 19.8 from a hexagonal extrusion and a cylinder which have been united using the **Union** tool.
2. *Click* on the **Offset Faces** tool icon in the **Solids Editing** toolbar (Fig.19.9). The command line shows:

Command:_solidedit
[prompts]:_face
[prompts]
[prompts]:_offset
Select faces or [Undo/Remove]: *pick* the bottom face of the 3D model **2 faces found**
Select faces or [Undo/Remove/All]: *enter* **r** *right-click*
Select faces or [Undo/Remove/All]: *pick* highlighted faces other than the bottom face **2 faces found, 1 removed**
Select faces or [Undo/Remove/All]: *right-click*
Specify the offset distance: *enter* **30** *right-click*

3. Repeat, offsetting the upper face of the cylinder by **50** and the right-hand face of the lower extrusion by **15**.

Fifth example – Taper Faces tool (Fig. 19.10)

1. Construct the 3D model as in the left-hand drawing of Fig. 19.10. Place in the **SW Isometric** view.
2. Call **Taper Faces**. The command line shows:

Command:_solidedit
[prompts]:_face
[prompts]
[prompts]:_taper

Fig. 19.8 Fourth example –
Offset Faces tool

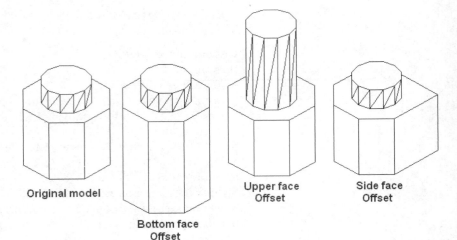

Original model

Bottom face
Offset

Upper face
Offset

Side face
Offset

Offset Faces

Fig. 19.9 The **Offset Faces** tool
icon from the **Solids Editing**
toolbar

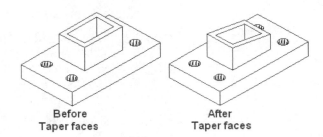

Before
Taper faces

After
Taper faces

Fig. 19.10 Fifth example – **Taper
Faces** tool

Select faces or [Undo/Remove]: *pick* the upper face of the base **2
faces found**
Select faces or [Undo/Remove/All]: *enter* **r** *right-click*
Select faces or [Undo/Remove/All]: *pick* highlighted faces other than
the upper face **2 faces found, 1 removed**
Select faces or [Undo/Remove/All]: *right-click*
Specify the base point: *pick* a point on left-hand edge of the face
Specify another point along the axis of tapering: *pick* a point on the
right-hand edge of the face
Specify the taper angle: *enter* **10** *right-click*

And the selected face tapers as indicated in the right-hand drawing.

Sixth example – Copy Faces tool (Fig. 19.13)

1. Construct a 3D model to the sizes as given in Fig. 19.12.
2. *Click* on the **Copy Faces** tool in the **Solids Editing** toolbar (Fig. 19.11).
   The command line shows:

   Command:_solidedit
   [prompts]:_face
   [prompts]

Copy Faces

Fig. 19.11 The **Copy Faces** tool
icon from the **Solids Editing**
toolbar

Fig. 19.12 Sixth example – **Copy Faces** tool – details of the 3D solid model

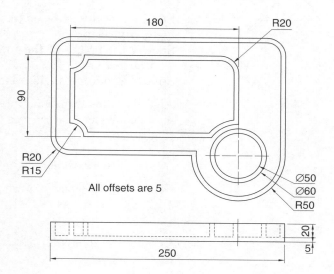

All offsets are 5

Fig. 19.13 Sixth example – **Copy Faces** tool

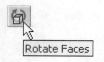

Fig. 19.14 The **Rotate Faces** tool icon from the **Solids Editing** toolbar

[prompts]:_copy
Select faces or [Undo/Remove]: *pick* the upper face of the solid model **2 faces found**
Select faces or [Undo/Remove/All]: *enter* **r** *right-click*
Select faces or [Undo/Remove/All]: *pick* highlighted face not to be copied **2 faces found, 1 removed**
Select faces or [Undo/Remove/All]: *right-click*
Specify a base point or displacement: *pick* anywhere on the high-lighted face
Specify a second point of displacement: *pick* a point some 50 units above the face

3. Add lights and a material to the 3D model and its copied face and render (Fig. 19.13).

Seventh example – Rotate Faces tool (Fig. 19.16)

1. Construct two cylinders to produce a 3D model (Fig. 19.15).
2. Call **Rotate Faces** from the **Solids Editing** toolbar (Fig. 19.14). The command line shows:

Command:_solidedit
[prompts]:_face
[prompts]
[prompts]:_rotate
Select faces or [Undo/Remove]: *pick* the left-hand face of the upper cylinder **2 faces found**
Select faces or [Undo/Remove/All]: *enter* **r** *right-click*
Select faces or [Undo/Remove/All]: *pick* unwanted faces **2 faces found, 1 removed**
Select faces or [Undo/Remove/All]: *right-click*

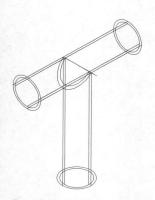

Fig. 19.15 Seventh example – **Rotate Faces** tool – the two cylinders

Fig. 19.16 Seventh example –
Rotate Faces tool

Specify an axis point or [Axis by object/View/Xaxis/Yaxis/Zaxis]:
enter **y** *right-click*
Specify the origin of the rotation: *pick* the upper point of the left-hand face
Specify a rotation angle: *enter* **10** *right-click*

3. Add lights and a material, place the 3D model in the **SW Isometric** view and render. The selected face is rotated as shown in the rendering (Fig. 19.16).

Eighth example – Color Faces tool (Fig. 19.18)

1. Construct a 3D model of a wheel to the sizes as shown in Fig. 19.17.
2. *Click* the **Color Faces** tool icon in the **Solids Editing** toolbar. The command line shows:

Command:_solidedit
[prompts]:_face
[prompts]
[prompts]:_color
Select faces or [Undo/Remove]: *pick* the inner face of the wheel **2 faces found**
Select faces or [Undo/Remove/All]: *enter* **r** *right-click*
Select faces or [Undo/Remove/All]: *pick* highlighted faces other than the required face **2 faces found, 1 removed**
Enter new color <ByLayer>: *enter* **1** (which is red) *right-click*

3. Add lights and a material to the edited 3D model and render (Fig. 19.18).

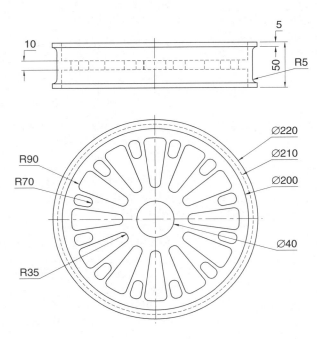

Fig. 19.17 Eighth example –
Color Faces tool – details of the
3D model

Fig. 19.18 Eighth example –
Color Faces tool

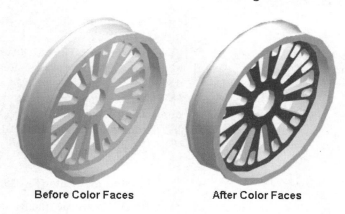

Before Color Faces   After Color Faces

Fig. 19.19 First example of 3D
models

Examples of more 3D models

These 3D models can mostly be constructed in the **acadiso.dwt** screen,
but some may require the **Limits** of the template to be increased to
594,420 (metric A3 sheet size). The descriptions of the stages needed to
construct these 3D models have been reduced from those given in earlier
pages, in the hope that readers have already acquired a reasonable skill in
the construction of such drawings.

First example (Fig. 19.19)

1. *FRONT* plane. Construct the three extrusions for the back panel and
   the two extruding panels to the details given in Fig. 19.20.
2. *WORLD* plane. Move the two panels to the front of the body and
   union the three extrusions. Construct the extrusions for the projecting
   parts holding the pin.

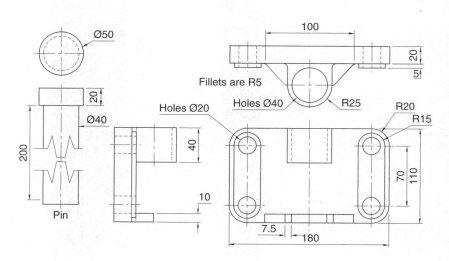

Fig. 19.20 First example of 3D
models – details of sizes and shapes

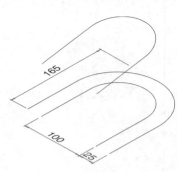

Fig. 19.21 Second example of 3D models – base plines

Fig. 19.22 Second example of 3D models

3. *FRONT* plane. Move the two extrusions into position and union them to the back.
4. *WORLD* plane. Construct two cylinders for the pin and its head.
5. *FRONT* plane. Move the head to the pin and union the two cylinders.
6. *WORLD* plane. Move the pin into its position in the holder. Add lights and materials.
7. **SW Isometric** view. Render. Adjust lighting and materials as necessary (Fig. 19.19).

Second example (Fig.19.22)

1. *WORLD*. Construct the two base plines in Fig. 19.21.
2. *FRONT*. Copy the inner pline **100** vertically upwards and rotate it **20°**.
3. **SW Isometric** view. Set the **SURFTAB** variables to suitable sizes and with **Edgesurf** form the surfaces as shown.
4. *WORLD*. Add lighting and a material to the model.
5. **SW Isometric** view. Render (Fig. 19.22).

Third example (Fig. 19.24)

1. *WORLD*. Fig. 19.23. Construct polyline outlines for the body extrusion and the solids of revolution for the two end parts. Extrude the body and subtract its hole and using the **Revolve** tool form the two end solids of revolution.
2. *RIGHT*. Move the two solids of revolution into their correct positions relative to the body and union the three parts. Construct a cylinder for the hole through the model.
3. *FRONT*. Move the cylinder to its correct position and subtract from the model.
4. *WORLD*. Add lighting and a material.
5. Render (Fig. 19.24).

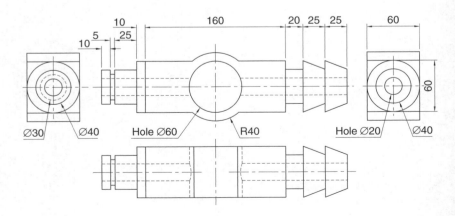

Fig. 19.23 Third example of 3D models – details

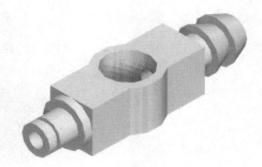

Fig. 19.24 Third example of 3D models

Fourth example (Fig. 19.27)

1. *RIGHT*. Construct two plines – Fig. 19.25.
2. *FRONT*. Move the smaller pline **200** to the right.
3. Rotate the two plines to slope at an angle of **10°** to the vertical.
4. *WORLD*. With the aid of the osnap **endpoint** construct two pline arcs joining the bottom ends of the two sloping plines. While in the *WORLD* plane add suitable lighting (Fig. 19.26).
5. **SW Isometric**. Set the **SURFTAB** variables to suitable sizes and with **Edgesurf** from a surface from the four plines.
6. Add a material and render (Fig. 19.27).

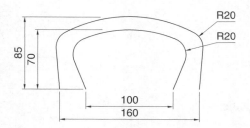

Fig. 19.25 Fourth example of 3D models – the two plines

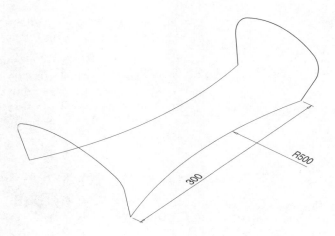

Fig. 19.26 Fourth example of 3D models – the four plines

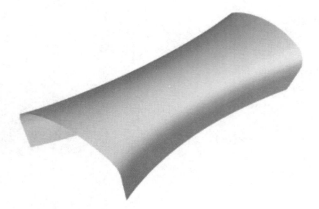

Fig. 19.27 Fourth example of 3D
models

Fifth example (Fig. 19.29)

1. *FRONT*. Construct the three plines needed for the extrusions of
   each part of the model (details Fig. 19.28). Extrude to the given
   heights. Subtract the hole from the **20** high extrusion.

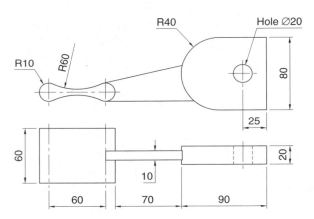

Fig. 19.28 Fifth example of 3D
models – details of shapes and
sizes

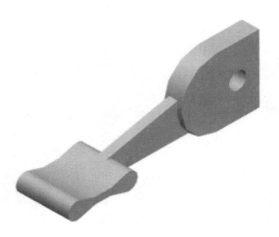

Fig. 19.29 Fifth example of 3D
models

2. *WORLD*. Move the **60** extrusion and the **10** extrusion into their correct positions relative to the **20** extrusion. With **Union** from a single 3D model from the three extrusions.
3. Add suitable lighting and a material to the model.
4. **SW Isometric**. Render (Fig. 19.29).

Sixth example (Fig. 19.30)

1. *FRONT*. Construct the polyline – left-hand drawing of Fig. 19.30.
2. With the **Revsurf** tool form a surface of revolution from the pline.
3. *WORLD*. Add suitable lighting and the material **GLASS**.
4. **SW Isometric**. Render (right-hand illustration of Fig. 19.30).

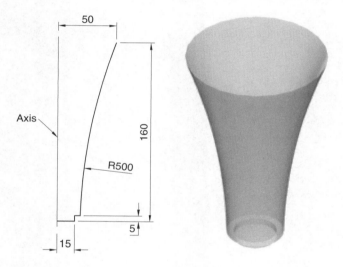

Fig. 19.30 Sixth example of 3D models

Exercises

1. Construct suitable polylines to sizes of your own discretion in order to form the two surfaces of the box shape shown in Fig. 19.31 with the aid of the **Rulesurf** tool.

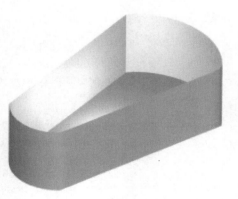

Fig. 19.31 Exercise 1 – first part

Fig. 19.32 Exercise 1 – second part

Add lighting and a material and render the surfaces so formed.

Construct another three **Edgesurf** surfaces to form a lid for the box. Place the surface in a position above the box, add a material and render (Fig. 19.32).

2. Working to the dimensions given in the orthographic projections of the three parts of this 3D model in Fig. 19.34, construct the assembled parts as shown in the rendered 3D model in Fig. 19.33.

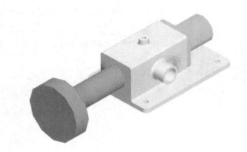

Fig. 19.33 Exercise 2

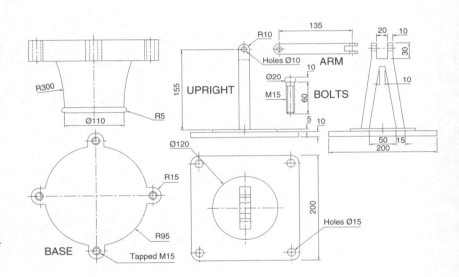

Fig. 19.34 Exercise 2 – details of shapes and sizes

Add suitable lighting and materials, place in one of the isometric viewing positions and render the model.

3. Construct polylines to the sizes and shapes as shown in Fig. 19.36. In order to form this surface, it is necessary to bisect the plan polylines and form the four edges needed for an edgesurf surface, then set the surftabs to suitable sizes, edgesurf the half surface. Then mirror the half surface to form the whole. Add lighting and a material and render (Fig. 19.35).

4. Construct the 3D model shown in the rendering in Fig. 19.38 from the details given in the parts drawing in Fig. 19.37.

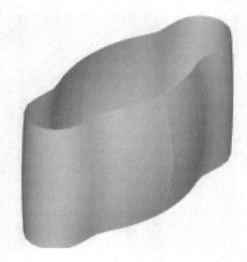

Fig. 19.35 Exercise 3

Fig. 19.36 Exercise 3 – plines

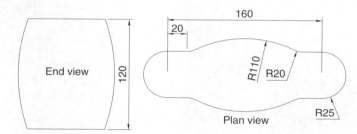

Fig. 19.37 Exercise 4 – the parts
drawing

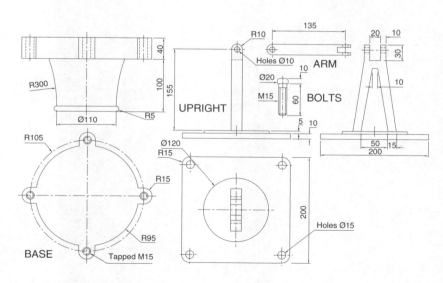

5. A more difficult exercise. A rendered 3D model of the two parts of an
assembly are shown in Fig. 19.39.

Working to the details given in the two orthographic projections in
Figs 19.40 and 19.41, construct the two parts of the 3D model, place

Fig. 19.38 Exercise 4

Fig. 19.39 Exercise 5

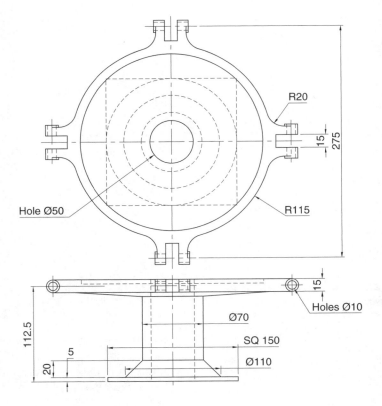

Fig. 19.40 Exercise 5 – the first
orthographic projection

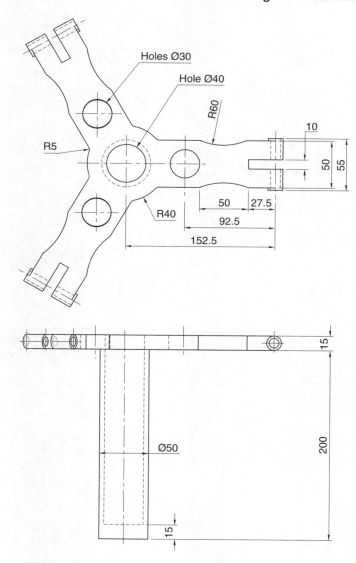

Fig. 19.41 Exercise 5 – the
second orthographic projection

them is suitable positions relative to each other, add lighting and materials and render the model.

Make sure the 3D model is saved to a suitable file name because it
is required for the completion of exercise 6.

6. Open the drawing saved in answer to exercise 5. Construct **2** links
to the details given in the two-view orthographic projections in
Fig. 19.42.

Copy the links each three times to produce **8** links in all and place
them in positions in the 3D model from exercise 5 as shown in the
rendering in Fig. 19.43.

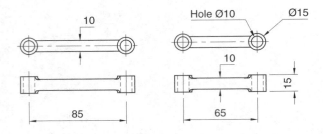

Fig. 19.42 Exercise 6 – the two links

Fillets are R2

Fig. 19.43 Exercise 6

CHAPTER 20

Other features of 3D modelling

Aims of this chapter

1. To give a further example of placing raster images in an AutoCAD drawing.
2. To give examples of the use of the **Setup Profile** tool to obtain profile-only drawings from 3D models.
3. To give examples of methods of printing or plotting not given in previous chapters.
4. To give examples of the use of polygonal viewports.

Raster images in AutoCAD drawings

Example – Raster image in a drawing (Fig. 20.5)

This example shows the raster file **rendering.bmp** of the 3D model constructed to the details given in the drawing in Fig. 20.1.

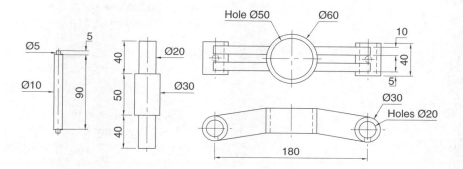

Fig. 20.1 Example – **Raster image in a drawing** – details

Raster images are graphics images such as those taken from files ending with the file extensions ***.bmp**, ***.pcx**, ***.tif** and the like. The types of graphics files which can be inserted into AutoCAD drawings can be seen by first *clicking* on **Raster Image** . . . in the **Insert** drop-down menu (Fig. 20.2), which brings the **Select Image File** dialog (Fig. 20.3) on screen. In the dialog *click* the arrow to the right of the **Files of type** field and the popup list which appears lists the types of graphics files which can be inserted into AutoCAD drawings. Such graphics files can be used to describe in 3D the details shown in 2D by a technical drawing.

Fig. 20.2 Selecting **Raster Image** . . . from the **Insert** drop-down menu

Fig. 20.3 The **Select Image File** dialog

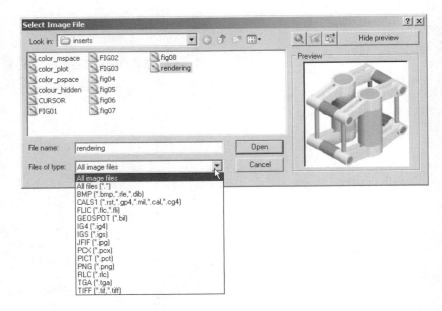

1. Construct the 3D model to the shapes and sizes given in Fig. 20.1, working in four layers each of a different colour.
2. Place in the **SW Isometric** view.
3. Shade the 3D model using **Gouraud Shading** from the **Shade** sub-menu of the **View** drop-down menu.
4. **Zoom** the shaded model to a suitable size and press the **Print Scr** key of the keyboard.
5. Open the Windows **Paint** application and *click* **Edit** in the menu bar, followed by another *click* on **Paste** in the drop-down menu. The whole AutoCAD screen which includes the Gouraud shaded 3D assembled mode appears.
6. *Click* the **Select** tool icon in the toolbar of **Paint** and window the 3D model. Then *click* **Copy** in the **Edit** drop-down menu.
7. *Click* **New** in the **File** drop-down menu, followed by a *click* on **No** in the warning window which appears.
8. *Click* **Paste** in the **Edit** drop-down menu. The shaded 3D model appears. *Click* **Save As** . . . from the **File** drop-down menu and save the bitmap to a suitable file name – in this example – **rendering.bmp**.
9. Open the orthographic projection drawing in AutoCAD.
10. Open the **Select Image File** and from the file list select **rendering.bmp** (Fig. 20.3). Another dialog (**Image**) opens (Fig. 20.4) showing the name of the raster image file. *Click* the **OK** button of the dialog and a series of prompts appear at the command line requesting position and scale of the image. *Enter* appropriate responses to these prompts and the image appears in position in the orthographic drawing (Fig. 20.5).

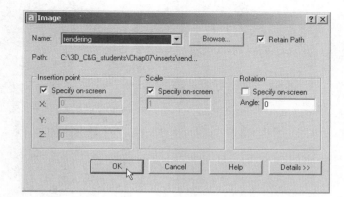

Fig. 20.4 Example – **Raster image in a drawing** – the **Image** dialog

Notes

1. It will normally be necessary to *enter* a scale in response to the prompt lines otherwise the raster image may appear very small on screen. If it does it can be zoomed anyway.
2. Place the image in position in the drawing area. In Fig. 20.5 the orthographic projections have been placed within a margin and a title block has been added.

Fig. 20.5 Example – **Raster image in a drawing**

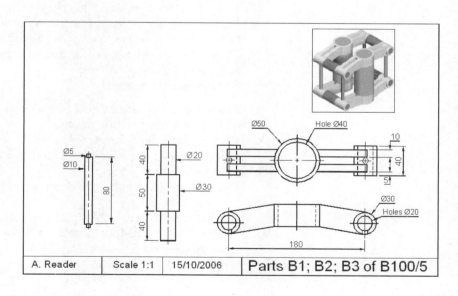

Fig. 20.6 The **Setup Profile** tool icon from the **Solids** toolbar

The Setup Profile tool

The **Setup Profile** tool is found in the **Solids** toolbar (Fig. 20.6). Its purpose is to create profile-only drawings from 3D solid drawings.

First example – Setup Profile (Fig. 20.10)

1. Construct a 3D solid model drawing to the details given in Fig. 20.7.

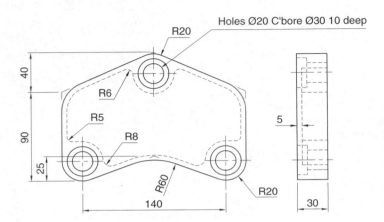

Fig. 20.7 First example – **Setup Profile** – orthographic projection

2. In the **UCS *FRONT*** plane copy the 3D model and mirror one of the copies to, in effect, turn it upside down (Fig. 20.8). Go back to the **UCS *WORLD*** plane.
3. Place the 3D model in one of the **3D Views** isometric views.
4. Change to a **PSpace** screen with a *click* on the **Layout1** tab.
5. *Click* the **Cancel** button of the **Page Setup** dialog which appears.
6. Either *enter* **ms** (MSpace) at the command line or *click* the **PAPER** button in the status bar to turn the **PSpace** screen into a **MSpace** screen.
7. Either *click* the **Setup Profile** tool icon in the **Solids** toolbar, or *enter* **solprof** at the command line.
   The command line shows:

   Command: *enter* **ms** (Model Space) *right-click*
   MSPACE Regenerating model.
   Command: *enter* **solprof** *right-click*
   Select objects: window the two models **2 found**
   Select objects: *right-click*
   Display hidden profile lines on separate layer [Yes/No] <Yes>:
      right-click
   Project profile lines onto a plane [Yes/No] <Y>: *right-click*
   Delete tangential edges [Yes/No] <Yes>: *right-click*
   2 solids selected
   Command:

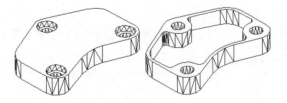

Fig. 20.8 First example – **Setup Profile** – the mirrored 3D models

Fig. 20.9 First example – **Setup Profile** – turn all layers off except **PV-2EO**

8. *Click* the arrow to the right of the **Layers** field and turn all layers off except that with a name commencing **PV** – in this example this is **PV-2E0** (Fig. 20.9). The 3D models appear in profile-only views (Fig. 20.10).

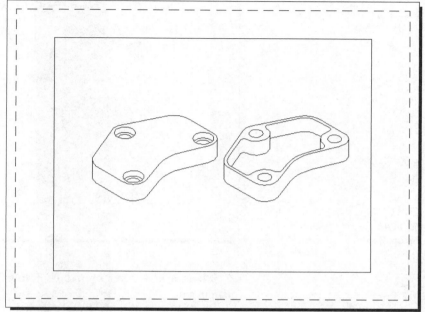

Fig. 20.10 First example – **Setup Profile**

Note

Profile-only views can only be created in **Layout** (**Pspace**) screens.

Second example – Setup Profile (Fig. 20.11)

Working to the same procedures as those shown for the first example, open the 3D model drawing of the first example (page 275) and create a profile-only view of the model. The only difference is that when selecting objects, the drawing must be windowed because there are **16** objects in the model.

Printing/Plotting

Hardcopy (prints or plots on paper) from a variety of AutoCAD drawings of 3D models can be obtained. Some of this variety has already been shown on pages 225–226 of Chapter 16.

First example – Printing/Plotting (Fig. 20.14)

If an attempt is made to print a multiple viewport screen with all viewport drawings appearing in the plot, only the current viewport will be printed. To print or plot all viewports:

1. Open a four-viewport screen of the assembled 3D model shown in the first example (page 275).

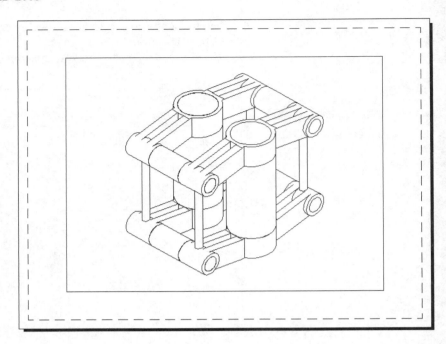

Fig. 20.11 Second example –
Setup Profile

2. Make a new layer **vports** of colour yellow. Make the layer **vports** current.
3. *Right-click* on the **Layout1** tab and *click* **Rename** in the menu which appears. Then in the **Rename Layout** dialog which comes to screen, *enter* a new name in the **Name** field. *Click* the **OK** button and the **Layout1** tab is renamed (Fig. 20.12).
4. *Click* the renamed **Layout** tab. The screen changes to a **PSpace** layout with the **Page Setup Layer1** dialog on top. In the dialog select the printer or plotter to be used by *clicking* the **Plot Device** tab and selecting the required printer/plotter from the **Name** popup list, followed by *clicking* the **Layout Settings** tab and selecting the required sheet size from the **Paper Size** popup list. *Click* the **OK** button of the dialog.

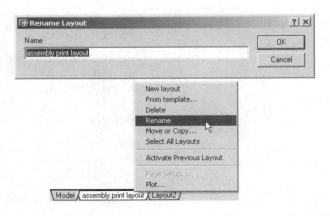

Fig. 20.12 First example –
Printing/Plotting – renaming
the **Layout1** tab

5. The Paper Space layout appears with the current viewport outlined in yellow (the colour of the **vports** layer). Using the **Erase** tool erase the viewport with a *click* on its boundary line. The viewport and its contents disappear.

6. At the command line:

Command: *enter* **mv** (Mview) *right-click*
MVIEW
Specify corner of viewport or [ON/OFF/Shadeplot/Lock/Object/ Polygonal/Restore/2/3/4]: *enter* **4** *right-click*
Specify first corner of viewport or [Fit] <Fit>: *right-click*
Regenerating model.
Command:

And four viewports appear with the 3D model drawing in each.

7. Make sure the variable **UCSFOLLOW** is off (**0**) in each viewport.

8. *Click* the **PAPER** button in the status bar to change the screen to **Model Space**.

9. Make each viewport in turn current with a *click* in its area and using the **3D Views** tools, make the top left viewport a **Top** view, the top right viewport a **Right** view, the bottom left viewport a **Front** view and the bottom right viewport a **SW Isometric** view.

10. In each viewport in turn **Zoom** the model drawing to a suitable size.

11. *Click* the **MODEL** tab to change the screen to a **Paper Space** layout.

12. *Click* the **Plot** tool icon in the **Standard** toolbar (Fig. 20.13). A **Plot** dialog appears.

13. Check that the printer/plotter is correct and the paper size is also correct.

14. *Click* the **Full Preview** button. The full preview of the plot appears (Fig. 20.14).

Fig. 20.13 The **Plot** tool icon from the **Standard** toolbar

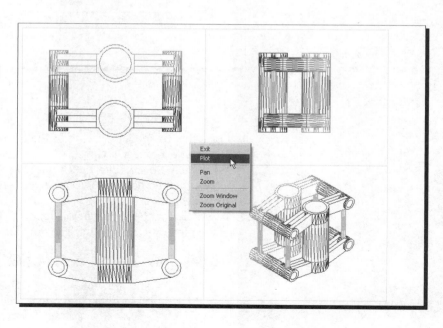

Fig. 20.14 First example – **Printing/Plotting**

15. *Right-click* anywhere in the drawing and *click* on **Plot** in the *right-click* menu which then appears.

16. The drawing plots (or prints).

Second example – Printing/Plotting (Fig. 20.15)

1. Open the orthographic drawing with its raster image shown in Fig. 20.5.

2. While still in **Model Space** *click* the **Plot** tool icon. The **Plot** dialog appears. Check that the required printer/plotter and paper size have been chosen.

3. *Click* the **Full Preview** button.

4. If satisfied with the preview (Fig. 20.15), *right-click* and in the menu which appears *click* the name **Plot**. The drawing plots.

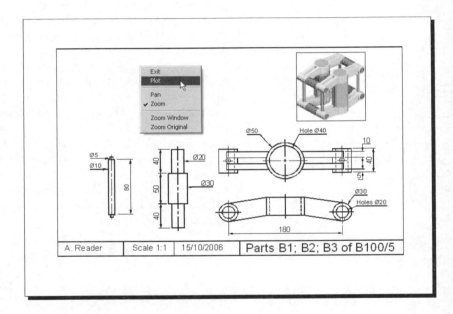

Fig. 20.15 Second example –
Printing/Plotting

Third example – Printing/Plotting (Fig. 20.16)

1. Open the 3D model drawing of the assembly shown in Fig. 20.14 in a single **SW Isometric** view.

2. While in **MSpace**, *click* the **Plot** tool icon. The **Plot** dialog appears.

3. Check that the plotter device and sheet sizes are correct. *Click* the **Full Preview** button.

4. If satisfied with the preview (Fig. 20.16), *right-click* and *click* on **Plot** in the menu which appears. The drawing plots.

Fourth example – Printing/Plotting (Fig. 20.17)

With the profile-only drawing on screen (Fig. 20.17), follow the procedure given with the third example to produce the required hardcopy.

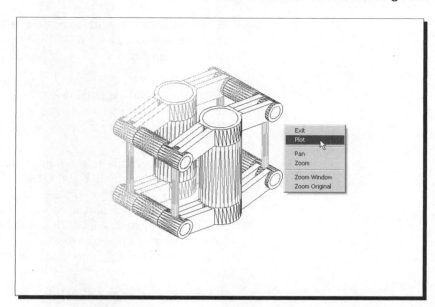

Fig. 20.16 Third example –
Printing/Plotting

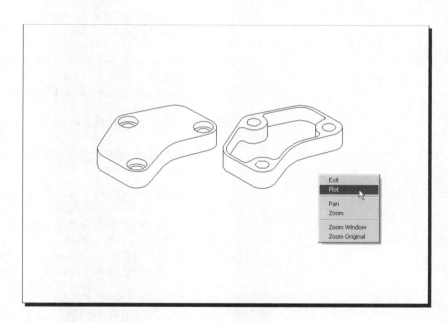

Fig. 20.17 Fourth example –
Printing/Plotting

Polygonal viewports (Fig. 20.18)

The example to illustrate the construction of polygonal viewports is based upon exercise 7 (page 289). When the 3D model for this assignment has been completed in **Model Space**:

1. Make a new layer **Yellow** of colour yellow and make this layer current.
2. *Click* the **Layout1** tab. Cancel the **Page Setup** dialog which appears.

3. Erase the viewport with a *click* on its bounding line. The outline and its contents are erased.

4. At the command line:

   Command: *enter* **mv** *right-click*
   [prompts]: *enter* **4** *right-click*
   [prompts]: *right-click*
   Regenerating model.
   Command:

   And the model appears in a four-viewport layout.

5. Make sure the variable **UCSFOLLOW** is **OFF** in each viewport.

6. *Click* the **PAPER** button in the status bar to turn it to **MODEL**. With a *click* in each viewport in turn and using the **3D Views** settings, set viewports in **Front**, **Right**, **Top** and **SW Isometric** views respectively.

7. **Zoom** each viewport to **All**.

8. *Click* the **MODEL** button to turn back to **PAPER**.

9. Enter **mv** at the command line which shows:

   Command: *enter* **mv** *right-click*
   MVIEW
   [prompts]: *enter* **p** (Polygonal) *right-click*
   Specify start point: In the top right viewport *pick* one corner for a square
   Specify next point or [Arc/Close/Length/Undo]: *pick* next corner for the square
   Specify next point or [Arc/Close/Length/Undo]: *pick* next corner for the square
   Specify next point or [Arc/Close/Length/Undo]: *enter* **c** (Close) *right-click*
   Regenerating model.
   Command:

   And a square viewport outline appears in the top right viewport within which is a copy of the model.

10. Repeat in each of the viewports with different shapes of polygonal viewport outlines (Fig. 20.18).

11. *Click* the **PAPER** button to change to **MODEL**.

12. In each of the polygonal viewports make a different isometric view. In the bottom right viewport change the view using the **3D Orbit** tool.

13. Turn the layer **Yellow** off. The viewport borders disappear.

14. The **Plot Preview** appears (Fig. 20.19). Plot the screen.

Exercises

1. Fig. 20.20 shows a polyline for each of the four objects from which the surface shown in the illustration of a Gouraud shaded surface was obtained. Construct the surface and shaded with **Gouraud shading** (Fig. 20.21).

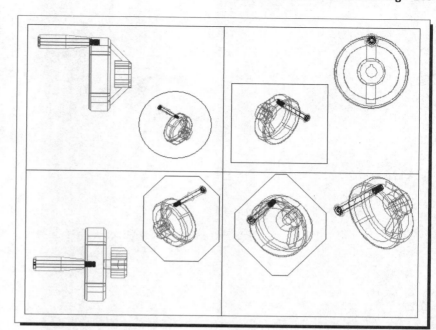

Fig. 20.18 **Polygonal viewports** –
plot preview

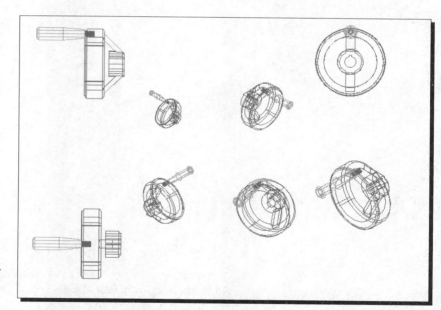

Fig. 20.19 **Polygonal viewports** –
plot preview after turning the layer
Yellow off

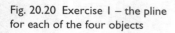

Fig. 20.20 Exercise 1 – the pline
for each of the four objects

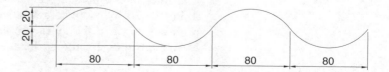

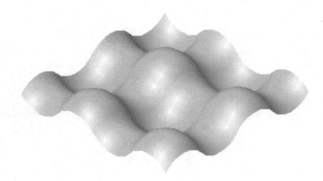

Fig. 20.21 Exercise 1

2. Working to the sizes given in Fig. 20.23, construct an assembled 3D model drawing of the spindle in its two holders and render (Fig. 20.22).

Fig. 20.22 Exercise 2

Fig. 20.23 Exercise 2 – details of shapes and sizes

Ø80   Ø100   Hole Ø80   R50   40   250   R50   R15   R120   90   40   20   170

3. A partial front view of a stand is shown by a profile-only drawing (Fig. 20.24). From the details given in the drawing in Fig. 20.25, construct a 3D model drawing of the stand. Using appropriate lighting and material render the 3D model which has been constructed.

4. Construct an assembled 3D model drawing, working to the details given in Fig. 20.26. When the drawing has been constructed, disassemble the parts as shown in the given profile-only drawing. Then construct the profile-only drawing in Fig. 20.27.

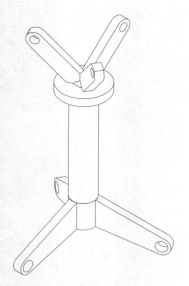

Fig. 20.24 Exercise 3 –
profile-only drawing

Fig. 20.25 Exercise 3

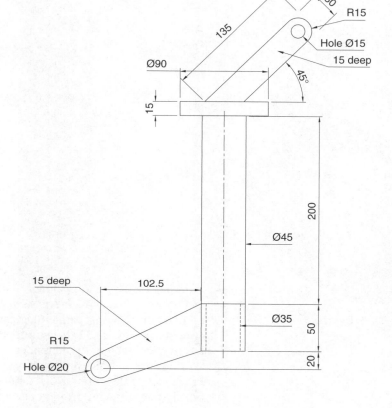

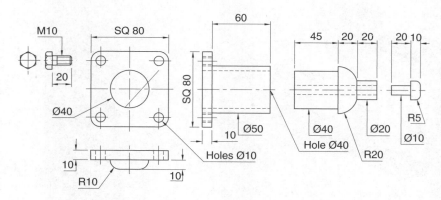

Fig. 20.26 Exercise 4 – details of
shapes and sizes

5. The surface model for this assignment was constructed from three
edgesurf surfaces, working to the suggested objects for the surface
as shown in Fig. 20.30. The third surface objects in each case are left
to your discretion. Fig. 20.28 gives the completed surface model.
Fig. 20.29 shows the three surfaces separated from each other.

Fig. 20.27 Exercise 4

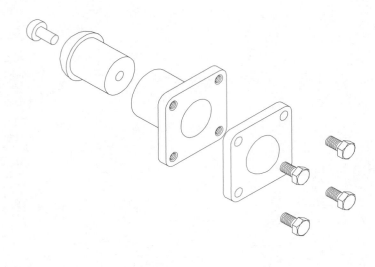

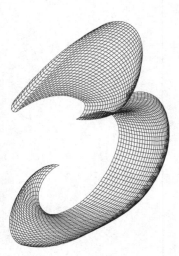

Fig. 20.28 Exercise 5

6. Working to the details shown in Fig. 20.31, construct a 3D model of the assembly, with the parts in their correct positions relative to each other. Then separate the parts as shown in the profile-only drawing in Fig. 20.32. When the 3D model is complete add suitable lighting and materials and render the result.

7. Working to the details shown in Fig. 20.33, construct an exploded 3D model of the parts of the handle and when completed construct a profile-only drawing from your model.

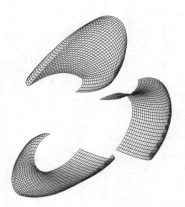

Fig. 20.29 Exercise 5 – surfaces separated

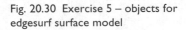

Fig. 20.30 Exercise 5 – objects for edgesurf surface model

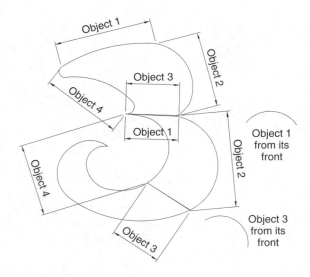

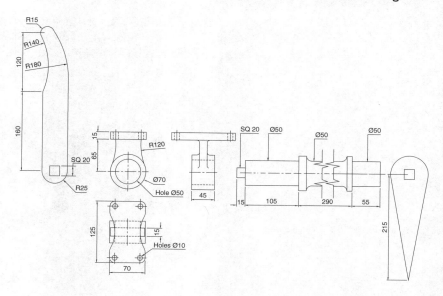

Fig. 20.31 Exercise 6 – details drawing

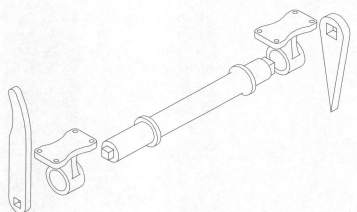

Fig. 20.32 Exercise 6 – profile-only drawing

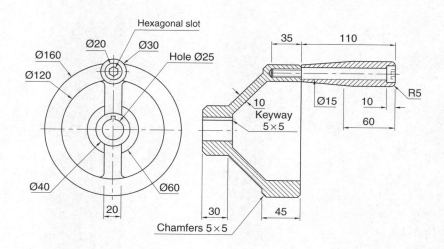

Fig. 20.33 Exercise 7 – details drawing

Two renderings of 3D models of the rotating handle are shown in Fig. 20.34 – one with its parts assembled and the other with the parts in an exploded position relative to each other.

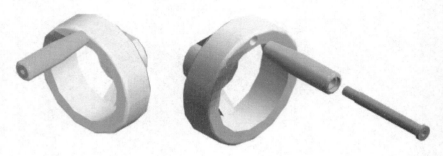

Fig. 20.34 Exercise 7 – renderings

CHAPTER 21

Internet tools

Aim of this chapter

The purpose of this chapter is to introduce the tools which are available in AutoCAD 2005, which make use of facilities available on the World Wide Web (www).

Emailing drawings

As with any other files which are composed of data, AutoCAD drawings can be sent by email as attachments. If a problem of security of the drawings is involved, they can be encapsulated with a password as the drawings are saved prior to being attached in an email. To encrypt a drawing with a password, *click* **Tools** in the **Save Drawing As** dialog and from the popup list which appears *click* **Security Options** . . . (Fig. 21.1). Then in the **Security Options** dialog which appears (Fig. 21.2) *enter* a password in the **Password or phrase to open this drawing** field, followed by a *click* on the **OK** button. The drawing then cannot be opened until the password

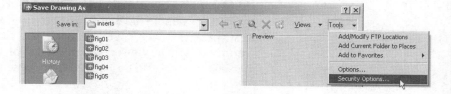

Fig. 21.1 Selecting **Security Options** . . . in the **Save Drawing As** dialog

Fig. 21.2 *Entering* a password in the **Security Options** dialog

is *entered* in the **Password** dialog (Fig. 21.3) which appears when an attempt is made to open the drawing by the person receiving the email.

There are many reasons why drawings may require to be password encapsulated in order to protect confidentiality of the contents of drawings.

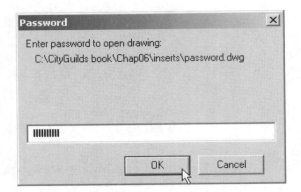

Fig. 21.3 The **Password** dialog appearing when a password-encrypted drawing is about to be opened

Example – creating a web page (Fig. 21.5)

To create a web page which includes AutoCAD drawings, *left-click* **Publish to Web** . . . from the **File** drop-down menu. A series of **Publish to Web** dialogs appear (Fig. 21.4). After making entries in the dialogs which come on screen after each **Next** button is *clicked*, the **Preview** of the resulting web page such as that shown in Fig. 21.5 (which can be saved as an ***.htm** file) appears. A *click* in any of the thumbnail views and an **Internet Explorer** page appears showing the selected drawing in full (Fig. 21.6).

In this example the thumbnails are from **DWF** files (**D**esign **W**eb **F**ormat files). If a web page includes DWF files they can be opened in the **Autodesk DWF Viewer** which is part of the AutoCAD 2005 software package. Fig. 21.7 shows a DWF file opened in the **DWF Viewer**.

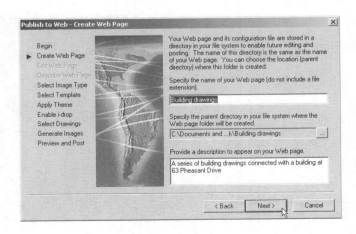

Fig. 21.4 One of the **Publish to Web** dialogs

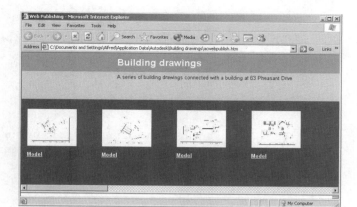

Fig. 21.5 A **Preview** of a web page

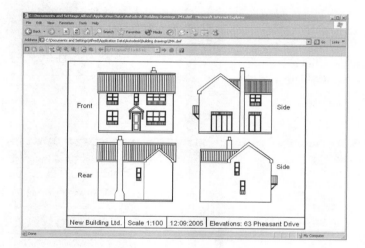

Fig. 21.6 The thumbnail in a **Preview** showing as a **Microsoft Explorer** page

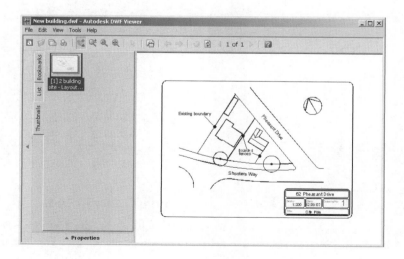

Fig. 21.7 A DWF file opened in the **Autodesk DWF Viewer**

Browsing the Web

There are several ways in which Autodesk web pages can be opened from AutoCAD 2005. A *click* on **Help** in the menu bar followed by another *click* on **Online Resources** in the drop-down menu and the sub-menu in Fig. 21.8 appears. Each of the items in this sub-menu brings an Autodesk web page to screen. As an example a *click* on the bottom item in the sub-menu and the web page Fig. 21.9 comes to screen. Selection of other items in the sub-menu brings up other Autodesk web pages.

Selecting **Browse the Web** from the **Web** toolbar (Fig. 21.10) also brings up an Autodesk web page.

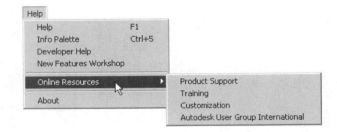

Fig. 21.8 The **Online Resources** sub-menu of the **Help** drop-down menu

Fig. 21.9 The **Autodesk Users Group** web page

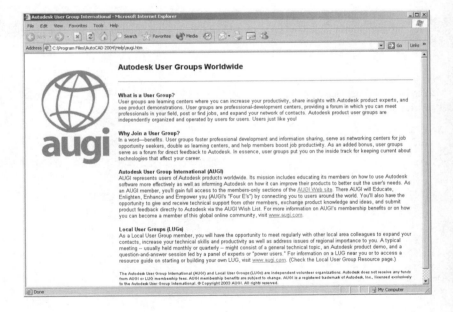

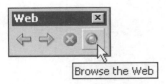

Fig. 21.10 Selecting **Browse the Web** from the **Web** toolbar

The eTransmit tool

Click **eTransmit . . .** in the **File** drop-down menu and the **Create Transmittal** dialog appears (Fig. 21.11). The transmittal shown in this example is

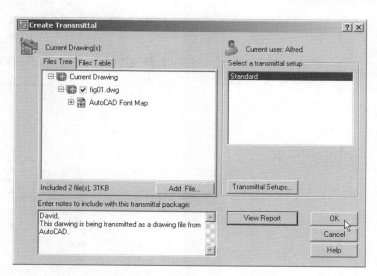

Fig. 21.11 The **Create Transmittal** dialog

the drawing on screen at the time. Fill in details as necessary and *click* the **OK** button and a **zip** file is formed from the drawing file (Fig. 21.12). This **zip** file is easier and quicker to email than the drawing file. The AutoCAD drawing can be obtained by unzipping the **zip** file at the receiving end.

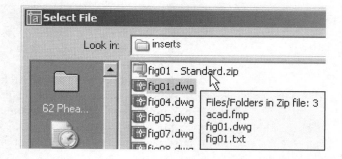

Fig. 21.12 The **zip** file created from the **Create Transmittal** dialog

CHAPTER 22

Design and AutoCAD 2005

Ten reasons for using AutoCAD

1. A CAD software package such as AutoCAD 2005 can be used to produce any form of technical drawing.
2. Technical drawings can be produced much more speedily using AutoCAD than when working manually – probably as much as ten times as quickly when used by skilled AutoCAD operators.
3. Drawing with AutoCAD is less tedious than drawing by hand – features such as hatching, lettering, adding notes, etc. are easier, quicker and indeed more accurate.
4. Drawings or parts of drawings can be moved, copied, scaled, rotated, mirrored and inserted into other drawings without having to redraw.
5. AutoCAD drawings can be saved to a file system without necessarily having to print the drawing. This can save the need for large drawing storage areas.
6. The same drawing or part of a drawing need never be drawn twice, because it can be copied or inserted into other drawings with ease. A basic rule when working with AutoCAD is: *Never draw the same feature twice*.
7. New details can be added to drawings or be changed within drawings without having to mechanically erase the old details.
8. Dimensions can be added to drawings with accuracy, reducing the possibility of making errors.
9. Drawings can be plotted or printed to any scale without having to redraw.
10. Drawings can be exchanged between computers and/or emailed around the world without having to physically send the drawing.

The place of AutoCAD 2005 in designing

The contents of this book are designed to help only those who have limited (or no) knowledge and skills of the construction of technical drawings using AutoCAD 2005. However it needs to be recognised that the impact of modern computing on the methods of designing in industry has been immense. Such features as analysis of stresses, shear forces, bending forces and the like can be carried out more quickly and accurately using computing methods. The storage of data connected with a design and the ability to recover the data speedily are carried out much easily using computing methods than prior to the introduction of computing.

AutoCAD 2005 can play an important part in the design process because technical drawings of all types are necessary for achieving well-designed artefacts whether it be an engineering component, a machine, a building, an electronics circuit or any other design project.

In particular, 2D drawings which can be constructed in AutoCAD 2005 are still of great value in modern industry. AutoCAD 2005 can also be used to produce excellent and accurate 3D models, which can be rendered to produce photographic like images of a suggested design. Although not dealt with in this book, data from 3D solid model drawings constructed in AutoCAD 2005 can be taken for use in computer aided machining (CAM).

At all stages in the design process either 2D or 3D drawings (or both) play an important part in aiding those engaged in designing to assist in assessing the results of their work at various stages. It is in the design process that drawings constructed in AutoCAD 2005 play an important part.

In the simplified design process chart shown in Fig. 22.1, an asterisk (*) has been shown against those features where the use of AutoCAD 2005 can be regarded as being of value.

The design chart (Fig. 22.1)

The simplified design chart in Fig. 22.1 shows the following features:

Design brief: A design brief is a necessary feature of the design process. It can be in the form of a statement, but it is usually much more. A design brief can be a written report which not only includes a statement made of the problem which the design is assumed to be solving, but includes preliminary notes and drawings describing difficulties which may be

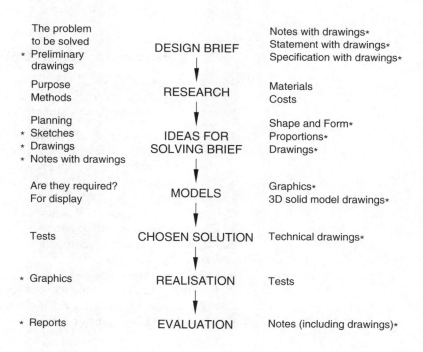

Fig. 22.1 A simplified design chart

encountered in solving the design and may include charts, drawings, costings, etc. to emphasise some of the needs in solving the problem for which the design is being made.

Research: The need to research the various problems which may arise when designing is often much more demanding than that shown in the chart (Fig. 22.1). For example the materials being used may require extensive research as to costing, stress analysis, electrical conductivity, difficulties in machining or in constructional techniques and other such features.

Ideas for solving the brief: This is where technical and other drawing and sketches play an important part in designing. It is only after research that designers can ensure the design brief will be fulfilled.

Models: These may be constructed models in materials representing the actual materials which have been chosen for the design, but in addition 3D solid model drawings, such as those which can be constructed in AutoCAD 2005, can be of value. Some models may also be made in the materials from which the final design is to be made so as to allow testing of the materials in the design situation.

Chosen solution: This is where the use of drawings constructed in AutoCAD 2005 are of great value. 2D and 3D drawings come into their own here. It is from such drawings that the final design will be manufactured.

Realisation: The design is made. There may be a need to manufacture a number of the designs in order to enable evaluation of the design to be fully assessed.

Evaluation: The manufactured design is tested in situations where it is liable to be placed in use. Evaluation will include reports and notes which could include drawings with suggestions for amendments to the working drawings from which the design was realised.

Enhancements in AutoCAD 2005

The enhancements of AutoCAD 2005 over the previous release AutoCAD 2004 include the following. It should be noted that details of many of these are not described in this introductory book.

1. The introduction of Sheet Sets, controlled by a **Sheet Set Manager**.
2. The introduction of Fields allowing name of current file, plot dates, object properties, etc. to be included in a drawing.
3. The introduction of tables which can be copied from Microsoft Excel or vice versa.
4. Tool palettes updated. Palettes can be made transparent if required allowing constructions to continue 'in front of' a transparent palette.
5. Layer management is more efficient in this new release using a **Layer Properties Manager**.
6. Plotting has been simplified.
7. **Etransmit** has been made more efficient.
8. **DWF** files can be used more efficiently.
9. An updated **DWF Viewer** is available.

System requirements for running AutoCAD 2005

Operating system: Windows XP Professional, Windows XP Home or Windows 2000.

Microsoft Internet Explorer 6.0.

Processor: Pentium III 800 MHz.

Ram: At least 128 MB.

Monitor screen: 1024 × 768 VGA with True Colour as a minimum.

Hard disk: A minimum of 300 MB.

APPENDIX A

Printing/Plotting

Introduction

Some suggestions for printing/plotting of AutoCAD drawings have already been given (pages 279 to 283). Plotters or printers can be selected from a wide range and are used for printing or plotting drawings constructed in AutoCAD 2005. The example given in this appendix has been from a print using one of the default printers connected to the computer I am using. However if another plotter or printer is connected to the computer, its driver can be set by first opening the Windows **Control Panel** and with a *double-click* on the **Autodesk Plotter Manager** icon the **Plotters** dialog appears (Fig. A.1).

Fig. A.1 The **Plotters** window

Double-click on the **Add-A-Plotter Wizard** icon and the first of the **Add Plotter – Plotter Model** series of dialogs appears (Fig. A.2). *Click* on any one of the names in the **Manufacturers** list and a selection can be made from the **Models** list associated with the chosen manufacturer's name.

Plots or prints from drawings constructed in AutoCAD 2005 can be made from either Model Space or Paper Space.

An example of a printout

1. Either select **Plot** . . . from the **File** drop-down menu (Fig. A.3) or *click* the **Plot** tool icon in the **Standard** toolbar (Fig. A.4). The **Plot** dialog appears (Fig. A.5).

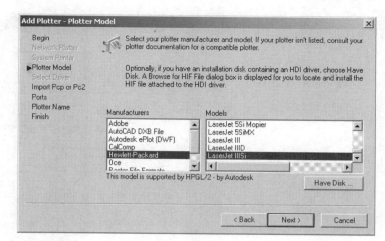

Fig. A.2 Selecting a printer or plotter from the **Add Plotter** dialog

Fig. A.3 Selecting **Plot** . . . from the **File** drop-down menu

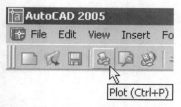

Fig. A.4 The **Plot** tool icon in the **Standard** toolbar

2. There are two parts in the **Plot** dialog. Fig. A.5 shows the main part. A *click* on the right-pointing arrow at the bottom right-hand corner of the dialog opens the second part (Fig. A.6).

3. Select an appropriate printer or plotter from the **Printer/Plotter** list. Then select the correct paper size from the **Paper size** popup list. Then select what is to be printed/plotted from the **What to plot** popup list. Finally *click* the **Preview** . . . button.

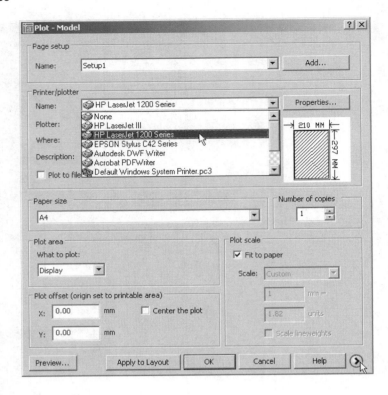

Fig. A.5 The main part of the
Plot dialog

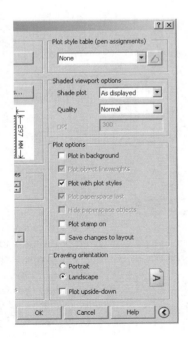

Fig. A.6 The second part of the
Plot dialog

4. A preview of the drawing to be printed/plotted appears (Fig. A.7). If satisfied with the preview, *right-click* and from the menu which appears *click* **Plot**. If not satisfied, *click* **Exit**. The preview disappears and the **Plot** dialog reappears. Make changes as required from an inspection of the preview and carry on in this manner until a plot can be made.

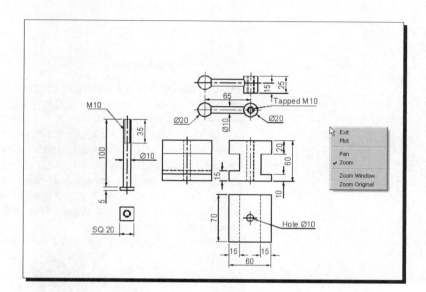

Fig. A.7 The **Plot Preview** window with its *right-click* menu

List of tools

Introduction

AutoCAD 2005 allows the use of over 300 tools. Some operators prefer using the word 'commands', although command as an alternative to tool is not in common use today. The majority of these tools are described in this list. Many of the tools described here have not been used in this book, because this book is an introductory text designed to initiate readers into the basic methods of using AutoCAD 2005. It is hoped the list will encourage readers to experiment with those tools not described in the book. The abbreviations for tools which can be abbreviated are included in brackets after the tool name. Tool names can be *entered* in upper or lower case.

A list of 2D tools is followed by a list of 3D tools. Internet tools are described at the end of this listing.

2D tools

About – Brings the **About AutoCAD** bitmap on screen
Adcenter – Brings the **DesignCenter** palette on screen
Appload – Brings the **Load/Unload Applications** dialog to screen
Arc (a) – Creates an arc
Area – States in square units the area selected from a number of points
Array (ar) – Creates **Rectangular** or **Polar** arrays in 2D
Ase – Brings the **dbConnect Manager** on screen
Attdef – Brings the **Attribute Definition** dialog on screen
Attedit – Allows editing of attributes from the Command line
Audit – Checks and fixes any errors in a drawing
Bhatch (h) – Brings the **Boundary Hatch** dialog on screen
Block – Brings the **Block Definition** dialog on screen
Bmake (b) – Brings the **Block Definition** dialog on screen
Bmpout – Brings the **Create Raster File** dialog
Boundary (bo) – Brings the **Boundary Creation** dialog on screen
Break (br) – Breaks an object into parts
Cal – Calculates mathematical expressions
Chamfer (cha) – Creates a chamfer between two entities
Chprop (ch) – Brings the **Properties** window on screen
Circle (c) – Creates a circle
Copy (co) – Creates a single or multiple copies of selected entities

Copyclip (Ctrl + C) – Copies a drawing, or part of a drawing, for inserting into a document from another application

Copylink – Forms a link between an AutoCAD drawing and its appearance in another application such as a word processing package

Customize – Brings the **Customize** dialog to screen, allowing the customisation of toolbars, palettes, etc.

Dblist – Creates a database list in a Text window for every entity in a drawing

Ddatext – Brings the **Attribute Extraction** dialog on screen

Ddattdef (at) – Brings the **Attribute Definition** dialog to screen

Ddatte (ate) – Edits individual attribute values

Ddcolor (col) – Brings the **Select Color** dialog on screen

Ddedit (ed) – The **Text Formatting** dialog box appears on selecting text

Ddim (d) – Brings the **Dimension Style Manager** dialog box on screen

Ddinsert (i) – Brings the **Insert** dialog on screen

Ddmodify – Brings the **Properties** window on screen

Ddosnap (os) – Brings the **Drafting Settings** dialog on screen

Ddptype – Brings the **Point Style** dialog on screen

Ddrmodes (rm) – Brings the **Drafting Settings** dialog on screen

Ddunits (un) – Brings the **Drawing Units** dialog on screen

Ddview (v) – Brings the **View** dialog on screen

Del – Allows a file (or any file) to be deleted

Dim – Starts a session of dimensioning

Dim1 – Allows the addition of a single dimension to a drawing

Dist (di) – Measures the distance between two points in coordinate units

Divide (div) – Divides an entity into equal parts

Donut (do) – Creates a donut

Dsviewer – Brings the **Aerial View** window on screen

Dtext (dt) – Creates dynamic text. Text appears in drawing area as it is *entered*

Dxbin – Brings the **Select DXB File** dialog on screen

Dxfin – Brings the **Select File** dialog on screen

Dxfout – Brings the **Save Drawing As** dialog on screen

Ellipse (el) – Creates an ellipse

Erase (e) – Erases selected entities from a drawing

Exit – Ends a drawing session and closes AutoCAD 2005

Explode (x) – Explodes a block or group into its various entities

Explorer – Brings the **Windows Explorer** on screen

Export (exp) – Brings the **Export Data** dialog on screen

Extend (ex) – To extend an entity to another

Fillet (f) – Creates a fillet between two entities

Filter – Brings the **Object Selection Filters** dialog on screen

Group (g) – Brings the **Object Grouping** dialog on screen

Hatch – Allows hatching by the *entry* responses to prompts

Hatchedit (he) – Allows editing of associative hatching

Help – Brings the **AutoCAD 2005 Help: User Documentation** dialog on screen

Hide (hi) – To hide hidden lines in 3D models

Id – Identifies a point on screen in coordinate units

Imageadjust (iad) – Allows adjustment of images

Imageattach (iat) – Brings the **Select Image File** dialog on screen

Imageclip – Allows clipping of images

Import – Brings the **Import File** dialog on screen

Insert (i) – Brings the **Inert** dialog on screen

Insertobj – Brings the **Insert Object** dialog on screen

Isoplane (Ctrl/E) – Sets the isoplane when constructing an isometric drawing

Layer (la) – Brings the **Layer Properties Manager** dialog on screen

Layout – Allows editing of layouts

Lengthen (len) – Lengthens an entity on screen

Limits – Sets the drawing limits in coordinate units

Line (l) – Creates a line

Linetype (lt) – Brings the **Linetype Manager** dialog on screen

List (li) – Lists in a text window the details of any entity or group of entities selected

Load – Brings the **Select Shape File** dialog on screen

Ltscale (lts) – Allows the linetype scale to be adjusted

Measure (me) – Allows measured intervals to be placed along entities

Menu – Brings the **Select Menu File** dialog on screen

Menuload – Brings the **Menu Customization** dialog on screen

Mirror (mi) – Creates an identical mirror image to selected entities

Mledit – Brings the **Multiline Edit Tools** dialog on screen

Mline (ml) – Creates a multiline

Mlstyle – Brings the **Multiline Styles** dialog on screen

Move (m) – Allows selected entities to be moved

Mslide – Brings the **Create Slide File** dialog on screen

Mspace (ms) – When in Paper Space changes to Model Space

Mtext (mt or t) – Brings the **Multiline Text Editor** on screen

Mview (mv) – To make settings of viewports in Paper Space

Mvsetup – Allows drawing specifications to be set up

New (Ctrl + N) – Brings the **Select template** dialog on screen

Notepad – For editing files from the Windows 95 **Notepad**

Offset (o) – Offsets selected entity by a stated distance

Oops – Cancels the effect of using **Erase**

Open – Brings the **Select File** dialog on screen

Options – Brings the **Options** dialog to screen

Ortho – Allows ortho to be set ON/OFF

Osnap (os) – Brings the **Drafting Settings** dialog on screen

Pagesetup – Brings either the **Page Setup Model** or **Page Setup – Layout1** dialog to screen for setting print/plot parameters

Pan (p) – *Drags* the AutoCAD 2005 drawing in any direction

Pbrush – Brings Windows **Paint** on screen

Pedit (pe) – Allows editing of polylines. One of the options is **Multiple**, allowing continuous editing of polylines without closing the command

Pline (pl) – Creates a polyline

Plot (Ctrl + P) – Brings the **Plot** dialog to screen

Point (po) – Allows a point to be placed on screen

Polygon (pol) – Creates a polygon

Polyline (pl) – Creates a polyline

Preferences (pr) – Brings the **Options** dialog on screen

Preview (pre) – Brings the print/plot preview box on screen

Properties – Brings the **Properties** palette on screen

Psfill – Allows polylines to be filled with patterns

Psout – Brings the **Create Postscript File** dialog on screen

Purge (pu) – Purges unwanted data from a drawing before saving to file

Qsave – Quicksave. Saves the drawing file to its current name in AutoCAD 2005 format

Quit – Ends a drawing session and closes down AutoCAD 2005

Ray – A construction line from a point

Recover – Brings the **Select File** dialog on screen to allow recovery of selected drawings as necessary

Rectang (rec) – Creates a pline rectangle

Redefine – If an AutoCAD command name has been turned off by **Undefine**, Redefine turns the command name back on

Redo – Cancels the last **Undo**

Redraw (r) – Redraws the contents of the AutoCAD 2005 drawing area

Redrawall (ra) – Redraws the whole of a drawing

Regen (re) – Regenerates the contents of the AutoCAD 2005 drawing area

Regenall (rea) – Regenerates the whole of a drawing

Region (reg) – Creates a region from an area within a boundary

Rename (ren) – Brings the **Rename** dialog on screen

Replay – Brings the **Replay** dialog on screen from which bitmap image files can be selected

Save (Ctrl + S) – Brings the **Save Drawing As** dialog box on screen

Saveas – Brings the **Save Drawing As** dialog box on screen

Saveimg – Brings the **Save Image** dialog on screen

Scale (sc) – Allows selected entities to be scaled in size – smaller or larger

Script (scr) – Brings the **Select Script File** dialog on screen

Setvar (set) – Can be used to bring a list of the settings of set variables into an AutoCAD Text window

Shape – Inserts an already loaded shape into a drawing

Shell – Allows MS-DOS commands to be *entered*

Sketch – Allows freehand sketching

Solid (so) – Creates a filled outline in triangular parts

Spell (sp) – Brings the **Check Spelling** dialog on screen

Spline (spl) – Creates a spline curve through selected points

Splinedit (spe) – Allows the editing of a spline curve

Stats – Brings the **Statistics** dialog on screen

Status – Shows the status (particularly memory use) in a Text window

Stretch (s) – Allows selected entities to be stretched

Style (st) – Brings the **Text Style** dialog on screen

Tablet (ta) – Allows a tablet to be used with a pointing device

Tbconfig – Brings the **Customize** dialog on screen to allow configuration of a toolbar

Text – Allows text from the command line to be *entered* into a drawing

Thickness (th) – Sets the thickness for the Elevation command

Tilemode – Allows settings to enable Paper Space

Tolerance – Brings the **Geometric Tolerance** dialog on screen

Toolbar (to) – Brings the **Toolbars** dialog on screen

Trim (tr) – Allows entities to be trimmed up to other entities

Type – Types the contents of a named file to screen

Undefine – Suppresses an AutoCAD command name

Undo (u) (Ctrl + Z) – Undoes the last action of a tool

View – Brings the **View** dialog on screen

Vplayer – Controls the visibility of layers in Paper Space

Vports – Brings the **Viewports** dialog on screen

Vslide – Brings the **Select Slide File** dialog on screen

Wblock (w) – Brings the **Create Drawing File** dialog on screen

Wmfin – Brings the **Import WMF File** dialog on screen

Wmfopts – Brings the **Import Options** dialog on screen

Wmfout – Brings the **Create WMF** dialog on screen

Xattach (xa) – Brings the **Select Reference File** dialog on screen

Xline – Creates a construction line

Xref (xr) – Brings the **Xref Manager** dialog on screen

Zoom (z) – Brings the zoom tool into action

3D tools

3darray – Creates an array of 3D models in 3D space

3dface (3f) – Creates a 3- or 4-sided 3D mesh behind which other features can be hidden

3dmesh – Creates a 3D mesh in 3D space

3dorbit (3do) – Allows manipulation of 3D models on screen

3dsin – Brings the **3D Studio File Import** dialog on screen

3dsout – Brings the **3D Studio Output File** dialog on screen

Align – Allows selected entities to be aligned to selected points in 3D space

Ameconvert – Converts AME solid models (from Release 12) into AutoCAD 2000 solid models

Box – Creates a 3D solid box

Cone – Creates a 3D model of a cone

Cylinder – Creates a 3D cylinder

Dducs (uc) – Brings the **UCS** dialog on screen

Edgesurf – Creates a 3D mesh surface from four adjoining edges

Extrude (ext) – Extrudes a closed polyline

Interfere – Creates an interference solid from selection of several solids

Intersect (in) – Creates an intersection solid from a group of two or more solids

Light – Brings the **Lights** dialog on screen

Matlib – Brings the **Materials Library** dialog on screen

Mirror3d – Mirrors 3D models in 3D space in selected directions

Mview (mv) – When in PSpace brings in MSpace objects

Pface – Allows the construction of a 3D mesh through a number of selected vertices

Plan – Allows a drawing in 3D space to be seen in plan (UCS World)

Pspace (ps) – Changes MSpace to PSpace

Render – Brings the **Render** dialog on screen

Revolve (rev) – Forms a solid of revolution from outlines

Revsurf – Creates a solid of revolution from a pline

Rmat – Brings the **Materials** dialog on screen

Rpref (rpr) – Brings the **Rendering Preferences** dialog on screen

Rulesurf – Creates a 3D mesh between two entities

Scene – Brings the **Scenes** dialog on screen

Section (sec) – Creates a section plane in a 3D model

Shade (sha) – Shades a selected 3D model

Slice (sl) – Allows a 3D model to be cut into several parts

Solprof – Creates a profile from a 3D solid model drawing

Sphere – Creates a 3D solid model sphere

Stlout – Saves a 3D model drawing in ASCII or binary format

Subtract (su) – Subtracts one 3D solid from another

Tabsurf – Creates a 3D solid from an outline and a direction vector

Torus (tor) – Allows a 3D torus to be created

UCS – Allows settings of the UCS plane

Union (uni) – Unites 3D solids into a single solid

Vpoint – Allows viewing positions to be set from x,y,z entries

Vports – Brings the **Viewports** dialog on screen

Wedge (we) – Creates a 3D solid in the shape of a wedge

Internet tools

Browser – Brings **www.autodesk.com** page on screen

Etransmit – Brings the **Create Transmittal** dialog to screen

Publish – Brings the **Publish to Web** dialog to screen

Some of the set variables

Introduction

AutoCAD 2005 is controlled by a large number of variables (over 440 in number), the settings of many of which are determined when making entries in dialogs. Others have to be set at the command line. Some are read-only variables which depend upon the configuration of AutoCAD 2005 when it was originally loaded into a computer (default values).

A list of those set variables follows which are of interest in that they often require setting by *entering* figures or letters at the command line. To set a variable, *enter* its name at the command line and respond to the prompts which arise.

To see all set variables, *enter* **set** (or **setvar**) at the command line:

> **Command:** *enter* **set** *right-click*
> **SETVAR Enter variable name or ?:** *enter* ?
> **Enter variable name to list <*>:** *right-click*

And a Text window opens showing a first window with a list of the first of the variables. To continue with the list, press the **Return** key when prompted and at each press of the **Return** key another window opens.

Some of the set variables

ANGDIR – Sets angle direction. **0** counterclockwise; **1** clockwise
APERTURE – Sets size of pick box in pixels
BLIPMODE – Set to **1** marker blips show; set to **0** no blips

Note

DIM variables – There are over 70 variables for setting dimensioning, but most are in any case set in the **Dimension Style** dialog or as dimensioning proceeds. However one series of the **Dim** variables may be of interest:
  DMBLOCK – Sets a name for the block drawn for an operator's own arrowheads. These are drawn in unit sizes and saved as required
  DIMBLK1 – Operator's arrowhead for first end of line
  DIMBLK2 – Operator's arrowhead for other end of line
DRAGMODE – Set to **0** no dragging; set to **1** dragging on; set to **2** automatic dragging
DRAG1 – Sets regeneration drag sampling. Initial value is 10

DRAG2 – Sets fast dragging regeneration rate. Initial value is 25

FILEDIA – Set to **0** disables **Open** and **Save As** dialogs; set to **1** enables these dialogs

FILLMODE – Set to **1** hatched areas are filled with hatching. Set to **0** hatched areas are not filled. Set to **0** plines are not filled

GRIPS – Set to **1** grips show; set to **0** grips do not show

MBUTTONPAN – Set to **0** no *right-click* menu with the Intellimouse; set to **1** Intellimouse *right-click* menu on

MIRRTEXT – Set to **0** text direction is retained; set to **1** text is mirrored

PELLIPSE – Set to **0** creates true ellipses; set to **1** polyline ellipses

PICKBOX – Sets selection pick box height in pixels

PICKDRAG – Set to **0** selection windows picked by two corners; set to **1** selection windows are dragged from corner to corner

RASTERPREVIEW – Set to **0** raster preview images not created with drawing; set to **1** preview images created

SHORTCUTMENU – For controlling how *right-click* menus show: **0** all disabled; **1** default menus only; **2** edit mode menus; **4** command mode menus; **8** command mode menus when options are currently available. Adding the figures enables more than one option

SURFTAB1 – Sets mesh density in the M direction for surfaces generated by the **Surfaces** tools

SURFTAB2 – Sets mesh density in the N direction for surfaces generated by the **Surfaces** tools

TEXTFILL – Set to **0** True Type text shows as outlines only; set to **1** True Type text is filled

TILEMODE – Set to **0** Paper Space enabled; set to **1** tiled viewports in Model Space

TOOLTIPS – Set to **0** no tool tips; set to **1** tool tips enabled

TPSTATE – Set to **0** and the Tool Palettes window is inactive; set to **1** and the Tool Palettes window is active

TRIMMODE – Set to **0** edges not trimmed when **Chamfer** and **Fillet** are used; set to **1** edges are trimmed

UCSFOLLOW – Set to **0** new UCS settings do not take effect; set to **1** UCS settings follow requested settings

UCSICON – Set **OFF** UCS icon does not show; set **ON** it shows

Computing terms

This glossary contains some of the more common computing terms.

Application – The name given to software packages which perform tasks such as word processing, Desktop Packaging, CAD, etc.

ASCII – The American National Standard Code for Information Interchange. A code which sets bits for characters used in computing.

Attribute – Text appearing in a drawing, sometimes linked to a block.

Autodesk – The American company which produces AutoCAD and other CAD software packages.

Baud rate – A measure of the rate at which a computer system can receive information (measured in bits per second).

Bios – Basic Input–Output System. The chip in a PC that controls the operations performed by the hardware (e.g. disks, screen, keyboard, etc.).

Bit – Short for binary digit. Binary is a form of mathematics that uses only two numbers: 0 and 1. Computers operate completely on binary mathematics.

Block – A group of objects or entities on screen that have been linked together to act as one unit.

Booting up – Starting up a computer to an operating level.

Bus – An electronic channel that allows the movement of data around a computer.

Byte – A sequence of 8 bits.

C – A computer programming language.

Cache – A section of memory (can be ROM or RAM) which holds data that is being frequently used. Speeds up the action of disks and applications.

CAD – Computer-aided design. The term should not be used to mean computer-aided drawing.

CAD/CAM – Computer-aided design and manufacturing.

CD-ROM – Computer disc read only memory. A disk system capable of storing several hundred MB of data – commonly 640 MB. Data can only be read from a CD-ROM, not written to it.

Chips – Pieces of silicon (usually) that drive computers and into which electronic circuits are embedded.

Command Line – In AutoCAD 2000, the Command Line is a window in which commands are *entered* from the keyboard and which contains the prompts and responses to commands.

Clock speed – Usually measured in MHz (Megahertz) – this is the measure of the speed at which a computer processor works.

Clone – Refers to a PC that functions in a way identical to the original IBM PC.

CMOS – Complimentary metal oxide semiconductor. Often found as battery-powered chips which control features such as the PC's clock-speed.

Communications – Describes the software and hardware that allow computers to communicate.

Compatibility – Generally used as a term for software or programs able to run on any computer that is an IBM clone.

Coprocessor – A processor chip in a computer which runs in tandem with the main processor chip and can deal with arithmetic involving many decimal points (floating-point arithmetic). Often used in CAD systems to speed up drawing operations.

CPU – Central processing unit. The chip that drives a PC.

Data – Information that is created, used or stored on a computer in digital form.

Database – A piece of software that can handle and organise large amounts of information.

DesignXML – An Autodesk **XML** computer program for representing design data over the Internet.

Dialog box – A window that appears on screen in which options may be presented to the user, or requires the user to input information requested by the current application.

Directories – The system in MS-DOS for organising files on disk. Could be compared with a folder (the directory) containing documents (the files).

Disks – Storage hardware for holding data (files, applications, etc.). There are many types: the most common are hard disks (for mass storage), floppy disks (less storage) and CD-ROMs (mass storage).

Display – The screen allowing an operator to see the results of his work at a computer.

DOS – Disk operating system. The software that allows the computer to access and organise stored data. MS-DOS (produced by the Microsoft Corporation) is the DOS most widely used in PCs.

DTP – Desktop publishing. DTP software allows for the combination of text and graphics into page layouts which may then be printed.

EMS – Expanded memory specification. RAM over and above the original limit of 640 KB RAM in the original IBM PC. PCs are now being built to take up to 128 (or even more) MB RAM.

Entity – A single feature in graphics being drawn on screen – a line, a circle, a point. Sometimes linked together in a block, when the block acts as an entity.

File – Collection of data held as an entity on a disk.

Fixed disk – A hard disk that cannot usually be easily removed from the computer as distinct from floppy disks which are designed to be easily removable.

Floppy disk – A removable disk that stores data in magnetic form. The actual disk is a thin circular sheet of plastic with a magnetic surface, hence the term 'floppy'. It usually has a firm plastic case.

Flyout – A number of tool icons which appear when a tool icon which shows a small arrow is selected from a toolbar.

Formatting – The process of preparing the magnetic surface of a disk to enable it to hold digital data.

ftp – File Transfer Protocol. An Internet protocol used to fetch a required resource from the World Wide Web (www) server.

Giga – Means 1 000 000 000. In computer memory terms 1000 MB (megabytes) – actually 1 073 741 824 bytes because there are 1024 bytes in a kilobyte (KB).

GUI – Graphical user interface. Describes software (such as Windows) which allows the user to control the computer by representing functions with icons and other graphical images.

Hardcopy – The result of printing (or plotting) text or graphics on to paper or card.

Hard disk – A disk, usually fixed in a computer, which rotates at high speed and will hold large amounts of data often up to 1 GB.

Hardware – The equipment used in computing: the computer itself, disks, printers, monitor, etc.

HTML – HyperText Markup Language. A computer language for setting up pages which can be sent via the Internet.

http – HyperText Transfer Protocol. An Internet protocol used to fetch a required resource from the World Wide Web (www) server.

Hz (hertz) – The measure of 1 cycle per second. In computing terms, often used in millions of Hertz – (megahertz or MHz) as a measure of the clock-speed.

IBM – International Business Machines. An American computer manufacturing company – the largest in the world.

Intel – An American company which manufactures the processing chips used in the majority of PCs.

Internet – A network of computers linked in a world wide system by telephone.

Joystick – A small control unit used mainly in computer games. Some CAD systems use a joystick to control drawing on screen.

Kilo – Means 1000. In computing terms 1 KB (kilobyte) is 1024 bytes.

LAN – Local area network. Describes a network that typically links PCs in an office by cable where distances between the PCs is small.

LED – Light emitting diode.

Library – A set of frequently used symbols, phrases or other data on disk, that can be easily accessed by the operator.

Light pen – Used as a stylus to point directly at a display screen sensitive to its use.

Memory – Any medium (such as RAM or ROM chips) that allows the computer to store data internally that can be instantly recalled.

Message box – A window containing a message to be acted on which appears in response when certain tools or commands are selected.

MHz – Megahertz – 1 000 000 Hz (cycles per second).

Microcomputer – A PC is a microcomputer; a minicomputer is much larger and a mainframe computer is larger still. With the increase in memory possible with a microcomputer, the term seems to be dropping out of use.

Microsoft – The American company which produces Windows and MS-DOS software.

MIPS – Millions of instructions per second. A measure of a computer's speed – it is not comparable with the clock-speed as measured in MHz, because a single instruction may take more than a cycle to perform.

Monitor – Display screen.

Mouse – A device for controlling the position of an on-screen cursor within a GUI such as Windows.

MS-DOS – Microsoft Disk Operating System.

Multitasking – A computer that can carry out more than one task at a time is said to be multitasking. For example in AutoCAD Windows printing can be carried out 'in the background', while a new drawing is being constructed.

Multi-user – A computer that may be used by more than one operator.

Networking – The joining together of a group of computers allowing them to share the same data and software applications. LANs and WANs are examples of the types of networks available.

Object – A term used in CAD to describe an entity or group of entities that have been linked together.

Operating System – Software, and in some cases hardware, which allows the user to operate applications, and to organise and use data stored on a computer.

PC – Personal computer. Should strictly only be used to refer to an IBM clone, but is now in general use.

Pixels – The individual dots on a monitor display.

Plotter – Produces hardcopy of, for instance, a drawing produced on a computer by moving a pen over a piece of paper or card.

Printer – There are many types of printer: dot-matrix, bubble-jet and laser are the most common. Allows material produced on a computer (graphics and text) to be output as hardcopy.

Processor – The operating chip of a PC. Usually a single chip, such as the Intel 80386, 80486 or Pentium chip.

Program – A set of instructions to the computer that have been designed to produce a given result.

RAM – Random access memory. Data stored in RAM is lost when the computer is switched off, unless previously saved to a disk.

RGB – Red, green, blue.

ROM – Read only memory. Refers to those chips from which the data stored can be read, but to which data cannot be written. The data on a ROM is not lost when a computer is switched off.

Scanner – Hardware capable of being passed over a document or drawing, reading the image into a computer.

Software – Refers to any program or application that is used and run on a computer.

SQL – Structured query language. A computer programming language for translating and transferring data between an application, such as AutoCAD, and a database.

Toolbar – Toolbars contain a number of icons, representing tools.

Tools – Tools are usually selected from icons appearing in toolbars. A tool represents a command.

Tooltip – When a tool is selected by a *left-click* on its icon, a small box appears (a Tool Tip) carrying the name of the tool.

UNIX – A multiuser, multitasking operating system (short for UNICS: uniplexed information and computing systems).

URL – Uniform Resource Locator.

URL address – It has three parts; for example, in the address http://www.autodesk.com/acaduser, **http://** describes the service; **www.autodesk.com** is the Internet address; **acaduser** is the location at the Interent address.

VDU – Visual display unit.

Vectors – Refers to entities in computer graphics which are defined by the coordinates of end points of each part of the entity.

VGA – Video graphics array. Screen displays with resolution of up to 640×480 pixels in 256 colours. SVGA (Super VGA) will allow resolutions of up to 1024×768 pixels.

Virtual memory – A system by which disk space is used as if it were RAM to allow the computer to function as if more physical RAM were present. It is used by AutoCAD (and other software), but can slow down a computer's operation.

WAN – Wide area network. A network of computers that are a large distance apart – often communicating by telephone.

Warning box – A window containing a warning or request which the user must respond to, which appears when certain circumstances are met or actions are made.

WIMP – Windows, icons, mice and pointers. A term used to describe some GUIs.

Window – An area of the computer screen within which applications such as word processors may be operated.

Workstation – Often used to refer to a multiuser PC, or other system, used for the purposes of CAD (or other applications).

WORM – Write once read many. An optical data storage system that allows blank optical disks to have data written onto them only once.

www – World Wide Web.

WYSIWYG – What you see is what you get. What is seen on the screen is what will be printed.

Index